AF767648

CELESTIAL IMAGES

With Compliments,
— Michael Mendillo
GRS '68 '71

CELESTIAL IMAGES

ANTIQUARIAN ASTRONOMICAL CHARTS AND MAPS FROM THE MENDILLO COLLECTION

★

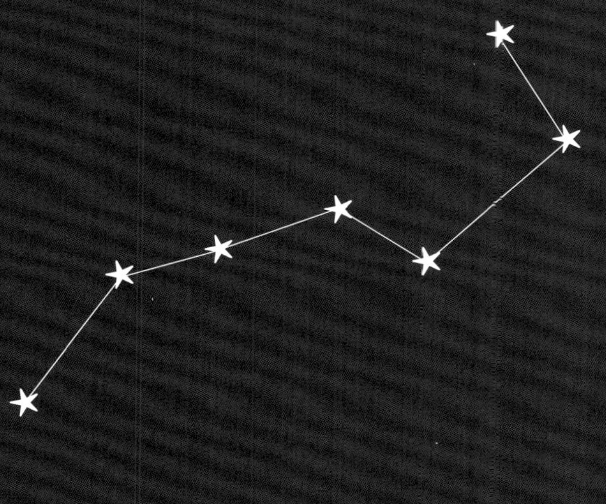

BOSTON UNIVERSITY ART GALLERY

BOSTON · 2005

COVER: Giovanni Maria Cassini (c. 1790). *Celestial Globe Gores: Sagitarrio, Capricorne, Aquario* from *Globo Celeste*, Rome, 1792. Engraving, 19¾″ x 13⅜″ (CAT. NO. 21)

FRONTISPIECE: Julius Schiller (d. 1627). *St. Joseph* from *Coelum Stellatum Christianum*, Augsburg, 1627. Engraving, 12⅛″ x 16⅛″ (CAT. NO. 72)

BOSTON UNIVERSITY ART GALLERY
855 Commonwealth Avenue
Boston, Massachusetts 02215
www.bu.edu/ART

Distributed by University of Washington Press
P.O. Box 50096
Seattle, Washington 98145-5096
www.washington.edu/uwpress

Printed in the United States of America

Library of Congress Control Number: 2004114872

ISBN: 1-881450-22-8

CONTENTS

FIGURE A. Allesandro Piccolomini (1508–1578), *The Constellation Orion* from *De le Stelle Fisse*, Venice, 1595 edition. Woodcut, 6.7″ x 5.5″. Courtesy of Michael Mendillo.

PREFACE

THIS PUBLICATION REPRESENTS the third manifestation of *Celestial Images* developed by the Boston University Art Gallery and we are pleased to carry on the tradition of this enduring interdisciplinary project that merges an interest in art and science. The original 1985 exhibition, *Celestial Images: Astronomical Charts from 1500 to 1900*, was a collaborative undertaking that resulted in an exhibition that traveled to the National Museum of American History, Smithsonian Institution, Washington, D.C.; the Williams College Museum of Art, Williamstown, Massachusetts; the Yale University Art Gallery, New Haven, Connecticut; and the New England Science Center, Worcester, Massachusetts. The exhibition was reprised in the year 2000 under the guidance of then Director John R. Stomberg, using objects from the Mendillo Collection only, and that exhibition opened at Boston University's 808 Gallery and traveled to the National Academy of Sciences, Washington, D.C. The 2000 exhibition checklist served as the basis for the current exhibition, and has been augmented to produce the eighty-two works in this traveling exhibition and corresponding catalogue. A successful exhibition catalogue also accompanied the original 1985 exhibition, and that publication has since gone out of print. We have continued to receive requests for that publication over the years, and those requests have stimulated a desire to revive the project once again in 2005, twenty years after its initial release.

The inspiration for the 1985 exhibition came from the collaborative efforts of then Director of the Boston University Art Gallery, Professor Patricia Hills, and Michael Mendillo, Professor of Astronomy at Boston University, and the Associate Dean of the Graduate School in 1985. Both the Art History Department and the Astronomy Department cooperated on the project. Patricia A. Johnston, then Acting Director of the Boston University Art Gallery, and her staff including Margaret Milhous, Susan Greendyke, Kathy Smith, and Walter King managed and provided additional support for the project.

Patricia M. Burnham, now Senior Lecturer in American Studies and Art History at The University of Texas at Austin, was guest curator of the exhibition for the Boston University Art Gallery in 1985, and as such was the driving force behind its initial success. With some alterations, we have reproduced here the original catalogue essays from the 1985 publication. For their continued assistance on this project and allowing us to reproduce their essays in this publication, we sincerely thank the authors including Patricia M. Burnham; Deborah Jean Warner, Curator, Physical Sciences Collection, National Museum of American History, Smithsonian Institution; and Samuel Y. Edgerton, Amos Lawrence Professor of Art, Williams College.

As the starting point for the extensive catalogue entries at the end of this publication, we have used many of the original catalogue entries in the 1985 publication. A research team including exhibition curator Patricia M. Burnham, exhibition consultant and Professor of Astronomy and History of Science at Harvard University Owen Gingerich, Professor Michael Mendillo, graduate assistant Margaret Milhous, and exhibition consultant Deborah Jean Warner wrote the entries, and we thank those authors for their tremendous efforts. We have omitted initials following each entry as they appeared in the 1985 publication, as those entries have been significantly edited and revised. We have also produced entirely new entries for many of the objects in the exhibition.

Michael Mendillo has been the guiding light and provided the passion for this project from the very beginning. The Mendillo Collection formed the core of the original 1985 exhibition, and completes the current exhibition. Michael Mendillo began collecting astronomical charts and maps in 1974, and as he explains in the introduction, his collection has served as a complement to his professional work in Astronomy. He is a vibrant and joyful spirit, and it has been a pleasure for all of us related to the exhibition to have the opportunity to work with him.

For their generous and continued support of this publication, we gratefully acknowledge Katherine T. O'Connor, Director, and Frances Heaton, Administrator, and the Humanities Foundation, College of Arts and Sciences and Graduate School of Arts and Sciences, at Boston University. We thank the Boston University Center for Space Physics for additional funding to facilitate the traveling exhibition. I wish to acknowledge the devoted efforts of the current staff at the Boston University Art Gallery, beginning with Assistant Director Rebekah Pierson. Graduate assistants Melissa Renn and Carrie Moore worked with multiple versions of exhibition checklists and catalogue text in order to generate the final versions. Undergraduate assistant Rebecca Love provided valuable proofreading skills and graduate assistant Joshua Buckno helped to facilitate the traveling exhibition. Professor Michael Mendillo would like to thank undergraduate Astronomy student Amanda Johnson for her research on cartographers and objects related to the exhibition, and colleagues Marina Galand and Sophie Laurent for their help with translations of French text. We also thank Paul Hoffmann and the entire staff of The Stinehour Press in Lunenburg, Vermont, for producing this fine publication.

Stacey McCarroll
Director & Curator
Boston University Art Gallery

INTRODUCTION

In the 21st century, posing the question "What do you see when you look up at the night sky?" would surely seem to be an odd thing to ask anyone in an urban environment. Even in remote locations, where light pollution is not extreme, a personal awareness of patterns in the darkness of space is not a priority of our times. Much earlier, the connections between the cosmos and everyday life were far more plentiful. As discussed in the three fine essays in this catalogue, the urges to name, to discover, to describe and to interpret celestial phenomena span human history. In previous millennia, the divisions between astrology, astronomy, mathematics and physics were essentially non-existent while, simultaneously, the artistic pursuits were partners with Natural Philosophy in showing, as Professor Edgerton points out in his essay, how we "see" the Universe and our place in it.

Visualization is a remarkably powerful tool in education, in the conduct of research, and for the portrayal of ideas and concepts, simple or complex. That is perhaps the major point that led me to begin collecting examples of how celestial "things" can be described. These range from maps of the sky's constellations, to schematics of the great theories for the movements of planets, to tutorial images of celestial phenomena deemed worthy of explanation. My interest in science occurred late in high school, and if President Kennedy had not excited me about flights to the Moon, I would have selected American History as my college major. In graduate school, I was fortunate to be in a department with Gerald Hawkins and Richard Berendzen as faculty members, extraordinary historians of astronomy, who offered humanistic breadth to my formal education and dissertation work in Space Physics. That fascination with astronomy, history, and visualization later came together when I was a dean in our Graduate School of Arts and Sciences and a young assistant professor, Patricia Hills, came to my office to argue for more graduate aid in Art History. She noticed two charts by Cellarius hanging on my office wall and, matter-of-factly, pointed to iconographic curiosities of gargoyle-like faces peering out from the borders around an image of Ptolemy's Earth-centered cosmology. I had not seen those little guys peering down at me for years, and the exhibition *Celestial Images* was born that afternoon, with that image (CAT. NO. 30), now one of ninety-three images in the collection on display.

I am often asked, "Where did you get all of these maps and charts?" Most came from antiquarian book and map shops in London, with others from Paris, Sydney, Ottawa, San Francisco, Padua, Munich, and Istanbul. Only a few were purchased locally in Boston. To most space scientists, the cities listed above are familiar as sites for national and international meetings during the past thirty years. One of the capabilities lost in professional mid-life is the ability to sit in convention centers listening intently to presentation after presentation that report incremental steps in science, and especially so during the visually-impaired years prior to PowerPoint. Rather than escape to a café or bar (in broad daylight), I often went to the section of town where the old books were sold. So, in a curious inversion of the very topic celebrated in these images, it was the frustration and boredom of science poorly portrayed that drove me to the streets in search of celestial wonders of bygone days. From the elegance of Jonathan Potter's gallery on New Bond Street in London, to the stalls of the extraordinary Grand Bazaar of Istanbul, the collection grew in ways as unpredictable as Koestler's famous "sleepwalker" description of the scientific method. The theme had a constancy centered on visual appeal, with gaps filled in when works by important cartographers appeared. In thirty years of collecting, I never went to a gallery looking for a particular item, and that added the element of discovery to an already exciting activity. From finding the large celestial globe gores by Coronelli

(CAT. NOS. 32–35) hanging in the window of a closed Chinese furniture store in San Francisco one night, to seeing a full set of Cassini constellation maps (CAT. NOS. 18–21) in a Sydney gallery specializing in prints of animals and birds, to trying to convince an antiquarian bookseller in Munich that while I would pay for her framed Doppelmayer's (CAT. NOS. 42–44) hanging on the walls, I wanted to give back the frames for easier transport home. The experiences were always exciting. For that Munich scenario, my lack of German language skills, when added to her lack of the tools needed to de-frame the three pieces, happily kept me there nearly as long as the scientific session I was playing hooky from.

In addition to my professorial mentors and colleagues who helped form my views and tastes, Dr. Howard Gotlieb of the Boston University's Special Collections (now the Howard Gotlieb Archival Research Center) played a pivotal role in the two crucial areas needed to create and to enjoy any type of collection: confidence and money. In discussions he urged me to collect what I liked and to hang them on the walls of my home to be enjoyed daily. When purchases of a complete volume and several individual prints by Cellarius appeared as a sudden possibility, he helped to arrange for financing and, later, he provided comfort from "buyer's regret" after the initial euphoria had passed. He and Katherine Kominis have continued to provide safe storage of my atlases, with access to students and scholars for research purposes. Katherine also facilitated my acquisition of the first volume of printed sky maps, Alessandro Piccolomini's *De le stelle fisse* (1540), completing a trio (with Bayer and Schiller) that is now my major focus of research.

Figure A (see p. 6) shows the stars that make up the constellation Orion in Piccolomini's book. CAT. NO. 5 gives Bayer's beautiful rendition of the image assigned to those stars by the ancients, while CAT. NO. 72 shows Schiller's version of "Christianizing the Heavens," with St. Joseph now representing those same stars. Note that the projection gets reversed from our terrestrial view (Orion's belt stars going from lower left to upper right in Bayer) to God's view from outside the celestial sphere ("seeing" those stars along the opposite diagonal) in Schiller. Of the three individuals involved, Piccolomini, Bayer and Schiller, it is surely Piccolomini who is the most interesting character (playwright, earthy poet, professor in Padua, bishop in Sienna); he was an early popularizer of science, a Carl Sagan of his day. Bayer and Schiller, on the other hand, were lawyers in Augsburg, with a passion for astronomy, and apparently not the Perry Masons of their day. As time permits, learning about them and the reasons behind Schiller's replacement scheme for all of the forty-eight constellations in Bayer's atlas keeps me on the road of discovery.

I also want to acknowledge Owen Gingerich from the Harvard-Smithsonian Center for Astrophysics. Owen helped in many ways prior to the first *Celestial Images* exhibition in 1985, and in the acquisition of my Bayer and Schiller atlases from Jonathan Potter's gallery. A bit more untraditional was his penchant to call saying, "I was in Paris (or Prague) last week and saw some Hevelius prints (or pages from Bode), and bought a few for you. Just give me a check when you come to pick them up!" If e-mail had existed a few decades ago, my collection might be larger and my credit in shambles.

The amount of effort and care needed to create an exhibition and catalogue is truly enormous, and the Boston University Art Gallery deserves special thanks for its management of *Celestial Images* over the past twenty years. The leadership, logistics and labor involved are, in my experience, comparable in scale to those associated with the preparation of an instrument for space flight. The Boston University Art Gallery is the exhibition's NASA—*Celestial Images* could not have been launched without it! Director & Curator Stacey McCarroll, Assistant Director Rebekah Pierson, past Directors John R. Stomberg and Patricia A. Johnston, and original Curator Patricia Burnham, together with their many students and staff members have, over the years, been joys to work with, and I am very grateful for that.

Finally, I want to thank my children, Johanna and Eric, who accompanied me on many trips to meetings when they were young and learned to curl up on the finely carpeted floors of galleries and antiquarian bookstores as Dad looked at old maps, in spite of having said earlier that the day would be devoted to more traditional methods of tourism.

MICHAEL MENDILLO
Professor of Astronomy
Boston University

CELESTIAL IMAGES: AN OVERVIEW

PATRICIA M. BURNHAM

THE PLACEMENT of fixed stars and the pattern of movement of the planets and other celestial bodies have been the focus of cosmological speculation, religious ritual, and practical research for thousands of years. For as many years, celestial bodies and stars have been the subject of artistic aspirations. Astronomical illustrations are among the earliest known visual images, from Bronze Age sun symbols to constellations found on Babylonian *kudurru* (boundary stones).

During the Middle Ages, astronomical lore chiefly appeared in manuscript form. Constellation figures, especially in the zodiac (because of its connection with astrology) constitute some of the most beautiful of medieval illuminations. Occasionally, there were representations of the heavenly spheres, geocentric in theory and simple in formulation. Around 1500, astronomical illustrations entered a new and lively phase. With the advent of printing, simultaneous issues of text and illustration in considerable numbers became possible. Before this, manuscripts had to be hand-copied, making accuracy difficult to achieve and restricting the spread of information.

The geographical explorations of the sixteenth century and the new discoveries in descriptive astronomy in the seventeenth century created a need for quick, accurate, and widespread dissemination of data. New stellar discoveries, planetary observations, and theoretical speculations called for a rapid means of communication that the printing process obligingly provided. The astronomical charts that resulted were beautiful as well as functional.

For almost four hundred years, astronomical charts combined a rich sense of form with practical utility. Often, the aesthetic additions included lavish decoration for decoration's sake, but even in those instances when diagrams were left relatively unadorned, ways were found to beautify them by using special typography, graphic design, and color. This kind of celestial cartography came to an end before 1900. Today's examples, although not lacking aesthetic properties of their own (incorporating unusual cartographic projections, advanced graphic design, and the glories of color printing), nevertheless differ from the kind of astronomical charts that began to be produced during the Renaissance and that are celebrated in this exhibition. The latter represented a balance of forces between science and art when the art of science and the science of art enjoyed a unique period of reciprocity. It would not be incorrect to refer to the printed charts produced from 1500 to 1900 as representing a "golden age" of celestial imagery.

Celestial Images begins with two different yet fundamental ways of describing the heavens: representations of planetary systems and charts of the stars. The former present a graphic view of cosmological theories. Star charts are essentially celestial cartography. A third category in the exhibition deals with the depiction of specific heavenly bodies and celestial phenomena.

Representations of planetary systems make a special contribution to our knowledge of intellectual and cultural history insofar as they reflect the dramatic changes in the relationship of the individual to the cosmos between 1500 and 1900. In ancient times, a distinction was made between the land-measurers (geometers) and those who speculated on the cosmos (philosophers). Perhaps a like distinction can be made between star charts, which are less theoretical, and planetary-systems charts, which often have serious philosophical ramifications.

From the time of the classical astronomer Ptolemy to the sixteenth century, a geocentric view of the heavens that posited the Earth as the central object in a small, simple universe, prevailed. The Polish monk Nicolaus Copernicus (1473–1543) revolutionized this view by devising a heliocentric model that placed the Sun at the center. The scholarly world reacted with acrimony to such a suggestion. Limited knowledge of the actual motions of the planets in addition to a literal reading of Scripture discouraged scholars from accepting Copernicus's hypothesis. Biblical texts frequently cited against Copernicus were Psalm 104, which referred to "the foundations of the Earth, that it should not be removed for ever," and Ecclesiastes 1:5, which described solar motions: "The Sun also ariseth, and the Sun goeth down, and hasteth to his place where he arose."

There was no rush to proclaim the new reforms in astronomical charts. For almost a century, the Earth-centered doctrine of Ptolemy competed with the newer ideas of Copernicus. Profusely ornamented charts recorded the argument, none more so than *Atlas Coelestis seu Harmonia Macrocosmica* by Andreas Cellarius (Amsterdam, 1660–1662) (CAT. NOS. 23–31). Galileo was tried by the Inquisition in 1633 for supporting Copernicus and, thirty years after the Galileo affair, Cellarius continued the attempt to reconcile the theories of Copernicus with Scripture and Church teaching. Little is known about Cellarius other than that he was a native of the Palatine in Germany, served as Rector of the Latin School at Hoorn in Northern Holland, and was known to be a learned scholar and mathematician.

Atlas Coelestis is a compendium of astronomical knowledge, paying respectful attention to the Danish astronomer Tycho Brahe and other philosophers of the heavens. The inclusion of Ptolemy is, by now, for historic purposes only. Although *Atlas Coelestis* has not been formally translated, informal analysis indicates that a special intensity of appeal and passion of language are reserved for the section on Copernicus. Cellarius's explanation for the statements in Scripture that seem to support geocentrism are disingenuous. In refuting the apparent allusion to motions of the Sun in Joshua 10:12 ("Sun, stand thou still upon Gibeon"), he claims that the Old Testament author was making a concession to lay understanding rather than defining a scientific truth. After all, the Sun *appeared* to be the body that moved. Echoing Galileo's apothegm, he pronounced: "It is not the scope of the Sacred Scripture to explain mathematics but to teach piety."[1] The splendid solar images of the Copernicus pages in the sumptuously illustrated volume proclaim his belief in the preeminence of the Sun among heavenly bodies.

Dynamic equilibrium between decoration and information was achieved in the pages of *Atlas Coelestis. Celestial Images* includes nine engravings from three editions that offer rich examples of the balance between science and art in the baroque era (CAT. NOS. 23–31). Not all examples of nonstellar phenomena are comparably embellished. Coming fifty years after Cellarius, de Fer's *Four Systems of Cosmology* shows cloud features without extraneous decoration (CAT. NO. 38). Doppelmayer's *Phaenomena* squeezes seven cosmological systems into a single chart, leaving room for merely a handful of angels and a large cartouche (CAT. NO. 43). The restraint of the Enlightenment informs Cassini's *Tavola Sferica* (CAT. NO. 22); Lapie offers nineteenth-century forthrightness, *Systèmes Planétaires* (CAT. NO. 59). By 1900, planetary information began to be transmitted by means of complex printed diagrams devoid of all ornamentation.

Such ornamentation seems almost to have enjoyed a metaphorical function in some of the early charts. In the Cellarius volume, for example, it celebrated the value of astronomical knowledge and the intellectual excitement aroused by new theories of the universe. After the storms over heliocentrism abated, new questions about the behavior of planets continued to be raised in astronomical charts (such as the occurrence and mobility of previously undiscovered planets), but with a decreasing ratio of embellishment to information. It was not merely that burgeoning information crowded out ornamentation, but that the expressive role of ornamentation itself ceased to be valued.

Printed star charts had a somewhat different history from the planetary systems images but suffered a similar fate. As far back as the Babylonians, people looked at the skies and imagined mythological creatures there. Classical writers gave the constellations their definitive identities. These iconographic conventions were retained during the Middle Ages, even though artistic technique declined. The astronomical chart was a direct beneficiary of the revival of classical art and learning that took place during the Renaissance. Ptolemy's text and other scientific treatises were translated; mathematics was improved; classical figure drawing was revived. A successful union of art and science was effected, lasting over three hundred years.

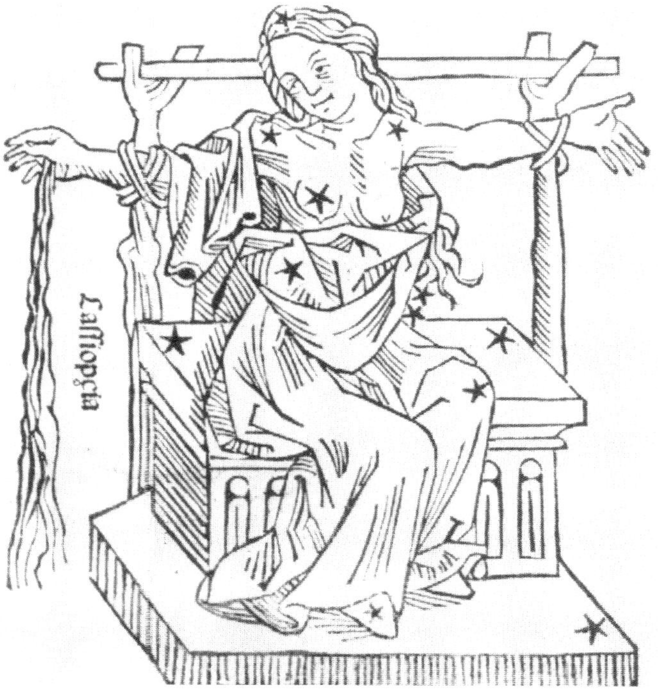

FIGURE B. Hyginus Astrologus (second century), *Andromeda* and *Cassiopeia* from *Clarissimi viri Igini Poeticon Astronomicon*, Venice, 1482 (Erhard Ratdoldt ed.). Woodcuts, 7″ x 6″ each. Courtesy of the Trustees of the Boston Public Library.

Only thirty years separate the woodcut illustrations in Erhard Ratdolt's edition of *Poeticon Astronomicon*, a fifteenth-century edition of star myths by the classical author, Hyginus (FIGURE B), and Albrecht Dürer's astronomical planispheres of 1515 (FIGURE C), yet they are cultural worlds apart. The Ratdolt edition of Hyginus is essentially medieval in character, whereas Dürer's star charts look toward the Renaissance in both their scientific and artistic aspects. Deborah Jean Warner, in her book, *The Sky Explored*, has pointed out that Dürer's printed planispheres were maps with coordinates and scales from which star positions could actually be plotted.[2] Dürer took his constellation designs both from earlier manuscript efforts in Germany and from the "new art" of the Italian Renaissance. Male nudes, such as Boötes and Hercules, were painstakingly cut from the block, careful hatching giving depth to the musculature and accuracy to the foreshortening. Andromeda was more crudely conceived, but plait-haired Virgo of the lilting gait, like an Alpine maenad, was the prototype for that constellation for years to come.

Although the remote sources for Dürer's figures were classical (filtered through late medieval German manuscripts and the Italian Renaissance), the figures differ considerably from those visible on an actual classical globe. The Farnese globe (a second century A.D. Roman copy of a third century B.C. Greek original) shows a standing Andromeda, a queenly Cassiopeia as a severe classical matron, a Sagittarius of Olympian calm, etc. Only the Farnese Virgo shows some similarity in pose to Dürer's, although she is clad in antique draperies rather than the late medieval German costume favored by Dürer.

Within the essential conservatism of constellation motifs, variations in representation often occurred. There was considerable disparity in technique, style, and perfection of form, as well as in the interpretation of subject. The principal outlines of figures had to be maintained in order for stars to be located easily, although this law was occasionally violated, causing confusion among astronomers. Variations of dress and accessories were permitted. As Warner points out, for example, Boötes was sometimes presented as a heroic Renaissance nude and sometimes dressed for a northern European winter.[3]

The major considerations—whether to depict the figures from the front, or geocentric viewpoint, or from the outside, that is, the "divine" point of view—troubled astronomers greatly.[4] Differences in pose

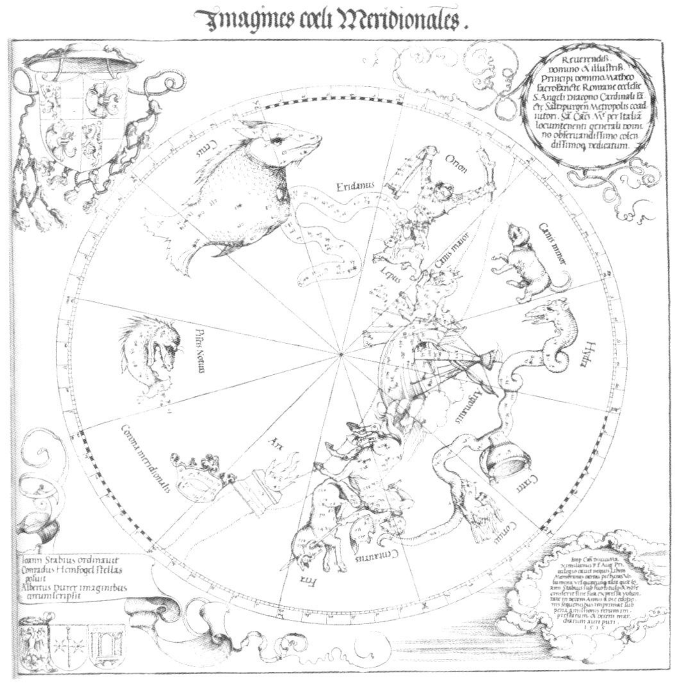

FIGURE C. Albrecht Dürer (1471–1528), *Imagines Coeli Meridionales* and *Imagines Coeli Septentrionales cum duodecim imaginibus zodiaci*, Nuremberg, 1515. Woodcuts, 17″ x 17″ each. Courtesy of the National Maritime Museum, London.

resulting from reversals, such as the representation of a frontally posed semi-seated Andromeda instead of a kneeling rear view, preserved the star locations but altered the iconography and resulted in peculiar awkwardnesses.

One of the most successful "outside" depictions of Virgo is that by Johannes Hevelius (CAT. NO. 52). Seen from the rear, with flowing tresses and right hand gesturing away from the body, she features exceptionally muscular wings that seem organically connected with her body and not merely added to the shoulders, as in so many other representations. Artists seemed to enjoy the special challenge of imagining the anatomy of winged creatures and treating it in such a way as to make it believable to the viewer. For a nineteenth-century school atlas in the United States, Elijah Burritt showed Virgo frontally, attired in an empire gown and holding the traditional sheaf of grain in her right hand (CAT. NO. 15).

Cassiopeia also provided artist-engravers with various opportunities not only to represent the female nude but also to interpret the popular type of the "humiliated woman." (Poseidon had consigned her to the heavens as a punishment; at certain times of the year she is upside-down, which explains why she is confined to her throne.) The motif of the woman bound is dealt with straightforwardly, voyeuristically, poetically, and playfully over the centuries. In 1482, Hyginus clearly emphasized the corporeal punishment accorded Cassiopeia and Andromeda. (FIGURE B)

Cassiopeia is shown completely naked in Dürer and emphatically so in Tycho Brahe's *Astronomiae*, legs crossed in a provocative way in accord with the conventions of the motif. Her flirtatious nature is manifested with baroque exuberance in Blaeu's gores. Here, sinuosity of line suggests the suppleness of flesh; the opulent curves of her throne rhyme with the curves of the body. Bayer's maiden (CAT. NO. 2) is modest by comparison. Burritt's *Atlas* shows a buxomly bare-breasted figure. Perhaps the classical origin of the constellation myth in combination with English art-historical sources provided Burritt with a rationale for allowing an undraped figure in the early nineteenth-century American classroom.

Usually, atlases with individual constellation pages gave more scope for figure drawing than single-sheet planispheres with multiple figures. The pages of Johann Bayer's atlas, for example, contain some of the most exquisitely drawn and colored human figures, animals, and still-life objects in all of celestial cartography. But such a mode of scientific communication eventually became anachronistic, superseded by the need to impart more stellar information than could be fitted into the constellation format. Bode's celestial atlas of 1801 was the swan song of great constellation design. As the accompanying chart shows, the number of newly discovered stars rose exponentially from the sixteenth century to the present (FIGURE D, p. 16). Today, the charted locations of more than one million stars are known. Only computers can cope with such numbers, which is why computer printouts and web pages have replaced printed star charts.

As with illustrations of planetary systems, star charts ceased to be works of art treasured by a literate public and became functional instruments used by specialists. The transformation of constellation figures to dots on computer printouts denotes the atomization of humankind; the heavens are no longer anthropomorphic images but depersonalized areas in the sky. *Celestial Images* permits us to re-enter the world of the past when science and art, like the twins of Gemini, walked hand in hand together. It allows us to repair to the realm of Urania for a while, where simple systems explained the universe, and people held friendly commerce with the skies. And yet, Samuel Edgerton, in his essay in this volume, would encourage us not to linger there, but to look forward to the future, and to consider the possibility that science and art may yet have something to say to each other, with art, once again, leading the way.

1. "Sacrae Scripturae scopus non est, homines Physicos, aut Mathematicos efficere, sed pios, & ad salutem, gratiam, & gloriam supernaturalem, atque aeternam idoneos & aptos reddere." Andreas Cellarius, *Atlas Coelestis seu Harmonia Macrocosmica* (Amsterdam, 1660–1662), p. 42.

2. Deborah Jean Warner, *The Sky Explored: Celestial Cartography, 1500–1800* (New York: A. R. Liss, 1979).

3. See Deborah Jean Warner's essay in this volume, p. 17.

4. Ibid., p. 19. Ptolemy described constellation figures as if seen from the Earth; others, as if seen from above the Earth.

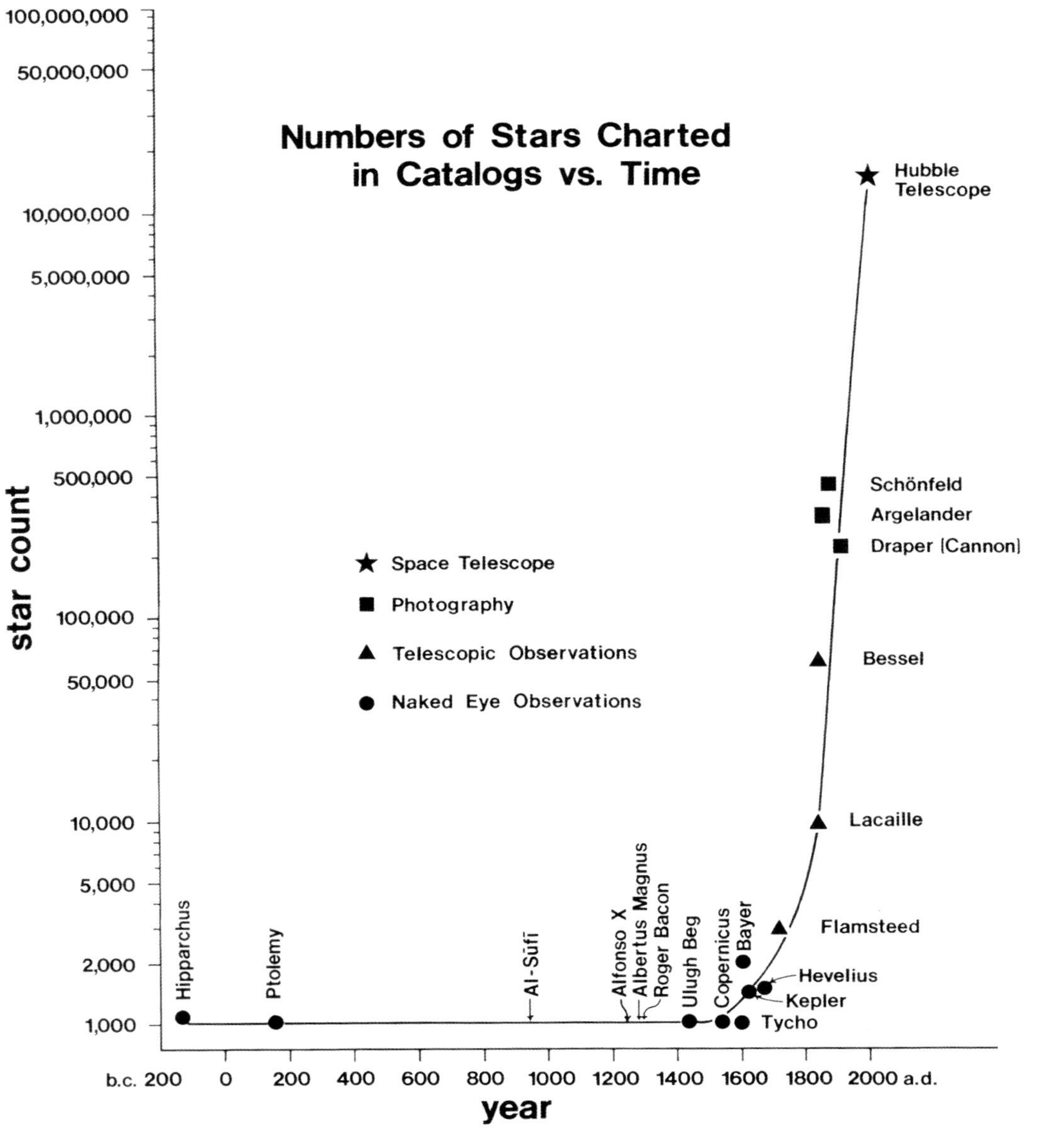

FIGURE D. *Number of Stars Charted in Catalogs vs. Time.* Courtesy of Michael Mendillo.

STAR CHARTS: THEIR LORE AND MEANING

DEBORAH JEAN WARNER

CONTEMPORARY ASTRONOMERS in well-equipped observatories can locate any celestial object from its coordinates and, conversely, they can easily determine the coordinates of newly discovered objects. Throughout most of history, however, people used constellations to find their way through the heavens, and they located new objects by measuring their distances from stars that were already charted. Thus for them star charts were indispensable tools, enjoyed for their beauty but depended on for their accuracy.

In Western tradition the most important star catalogue was that compiled by Ptolemy of Alexandria and included in the *Almagest* written around 150 A.D. The Ptolemaic catalogue lists the positions and magnitudes of 1028 stars (1025 plus 3 duplicates) visible from the Mediterranean and bright enough to be distinguished with the naked eye. Following ancient tradition Ptolemy grouped the stars into 48 constellations: 12 in the zodiac, 21 to the north, and 15 to the south. Although the Ptolemaic catalogue is far from complete, it defined the stellar universe for more than 1400 years (see FIGURE D, p. 16). During this time Islamic and European astronomers limited their observations to the Ptolemaic stars. The only changes they made in their description of the stellar sphere were to increase longitudes in accordance with precession (the 26,000-year passage of the vernal equinox around the ecliptic) and, occasionally, to observe latitudes and longitudes anew.

The Islamic globe in the collection of the National Museum of American History reflects the importance of the Ptolemaic tradition. Although oriental in style, the classical constellations are clearly identifiable. All the stars, except for a very few faint ones noted by Bedouins, are Ptolemaic and they are located within the constellation figures as described by Ptolemy. The star positions, however, are not strictly Ptolemaic, but derive from the version of the Ptolemaic catalogue compiled by Ulugh Beg at his observatory near Samarkand in the fifteenth century. Except for a few obviously misplaced stars, all the stellar latitudes on the globe are within 30 minutes of Ulugh Beg's values, and most are within 15 minutes. The longitudes are, on average, those of Ulugh Beg increased by 2 degrees 27 minutes. Thus, although this globe is undated, to the extent we can read its star positions we can assume it was meant to be correct for the year 1608. The two Islamic paintings shown in *Celestial Images* are again classical in their scientific content, but clearly Eastern in their iconographic style (CAT. NOS. 77 & 78).

Globes were the primary form of celestial map used throughout antiquity and the Middle Ages. Flat maps were known but most were simply illustrations from which neither positions nor distances could be read. The earliest extant flat celestial maps were drawn in Germany in the early fifteenth century, inspired in part by the recent recovery of Ptolemy's *Geography*. The oldest printed celestial maps derive from this tradition (See FIGURE C in Burnham's essay, p. 14). These very beautiful planispheres are usually attributed to the artist Albrecht Dürer but, like most maps, they resulted from the collaboration of several men. In this case Johann Stabius laid down the polar stereographic projection and drew the coordinates, Conrad Heinfogel positioned the stars, while Dürer drew the constellation figures and cut the wood blocks.

As an artist Dürer aimed to strip away the various Islamic influences, to recapture the essence of the constellations as they were known in classical antiquity. Accordingly, his human constellations are well-formed and muscular. Most are nude and their classical attributes have been restored. Perseus, for instance, holds Caput Medusa rather than the simple ghoulish head of Islamic tradition. This development of Renaissance constellation figures began with German artists in the mid-fourteenth century, but it reached its culmination with the work of Dürer.

The accuracy of any star map is a function of the catalogue on which it is based. Careful study of the Dürer planishperes has revealed many more errors than might be expected on such apparently precise maps. The best explanation seems to be that the errors originated not with Dürer and his colleagues, but in the astronomical authorities on which they relied. About 90 percent of the stars on the Dürer maps are placed within 1 degree of their Ptolemaic positions, while the remaining stars are badly misplaced. We cannot be certain of the immediate or ultimate source of the confusion, but we can identify several manuscript copies of the Ptolemaic catalogue with which Dürer and his colleagues might have been familiar and which have the same erroneous positions. A second problem concerns the epoch of the maps, the date for which they were meant to be correct. From a modern viewpoint the longitudes on these maps—Ptolemy plus 19 degrees 40 minutes—are correct for 1440 A.D. ±17½ years. Yet the maps are dated 1515. To find the correct contemporary longitudes Heinfogel probably relied on the *Alfonsine Tables*. These astronomical texts and tables had been compiled in the mid-thirteenth century under the sponsorship of King Alfonso X of Castile. They were first printed in 1483 and, although erroneous in some parts, they were in fact the best astronomical tables generally available at the time. The longitudinal difference between the stars on the planispheres and those listed in the *Alfonsine Tables*, 2 degrees 32 minutes, is just the value predicted for the end of 1502 by the Alfonsine trepidational theory. Ironically, Dürer lived in a house formerly owned by Bernard Walther, a Nuremberg astronomer aware of the problems with the Alfonsine values, who had spent more than a quarter of a century determining the star positions anew from his own observations.

Dürer, Heinfogel, and Stabius lived and worked in a Ptolemaic universe. But that universe was soon to change as astronomers began charting northern stars overlooked by Ptolemy, and voyaging into southern oceans to fill the southern circumpolar regions he could not see. As the number of known stars increased, astronomers and cartographers invented new constellations to encompass them. They also developed new cartographic projections and adopted various aesthetic and iconographic conventions. On Johannes Honter's planispheres of 1532 five of the male figures are dressed in contemporary fashions, with belted tunics and flamboyant hats. On his celestial globe of 1551 Gerard Mercator presented the constellation figures in Roman garb. At the end of the century Willem Blaeu depicted Argo as a Dutch ship of the period, and Boötes as dressed for a northern European winter.

For his *Uranometria* of 1603 Johann Bayer used poses and conventions, which had been popular in mythological works of the late fifteenth century (CAT. NOS. 2–5). To most astronomers Bayer is known for having introduced the stellar nomenclature—Greek letters identifying the brighter stars, Roman letters the fainter (and, for the most part, the order of the alphabets corresponds with decreasing brightness)—which, in time, replaced the Ptolemaic numbers. Yet, this convention had been introduced earlier by Piccolomini in 1540 (see FIGURE A, p. 6).

In the northern skies Bayer charted the Ptolemaic stars and several hundred others recently observed by the Danish astronomer Tycho Brahe. Like Brahe, Bayer described each of these stars as belonging to a Ptolemaic constellation, even though many of them fell outside the constellation figures. In the southern polar region Bayer charted stars observed by a Dutch navigator, Pietr Dirkz Keyser. These were grouped into twelve new constellations: Indus (an American Indian), Triangulum, and ten animals, some of which were native to the New World.

Two other non-Ptolemaic constellations in the *Uranometria*—Crux (the Southern Cross) and Columba (the dove sent by Noah to find dry land)—reflect the trend toward a Judeo-Christianization of the heavens. Crux, formed by Iberian navigators in the fifteenth century, and Columba, introduced by a Dutch geographer in the late sixteenth century, are still in use, as are Camelopardalis (the animal on which Rebecca rode into Canaan) and Monoceros (which appears at least four distinct times in the Old Testament). Other biblical constellations fell eventually into disuse, as did

Wilhelm Schickhard's explanation of the biblical significance of the classical groups.

The most thorough biblical reconstruction of the heavens was accomplished by Julius Schiller, who transformed the signs of the zodiac into the twelve apostles, the northern constellations into figures of the New Testament, and the southern constellations into figures of the Old. Schiller's religious revision never gained wide acceptance. From an astronomical point of view, however, his *Coelum Stellatum Christianum* of 1627 was the best set of celestial maps available in the middle decades of the seventeenth century (CAT. NOS. 70–73). Schiller was not an astronomer—like Bayer, he was a lawyer in Augsburg—but he relied on the best astronomical information of his day. In Andromeda, for instance, he included the great nebula now known as M31. Although Islamic texts had reported Bedouin sightings of this nebula, most European astronomers were unaware of it until 1612, when Simon Marius discovered it through a telescope.

Some celestial maps depict the sky as seen from the Earth. Others show the sky as if it were a globe viewed from the outside. Ptolemy adopted an external viewpoint and described the constellations as facing in toward the Earth. Dürer, who followed Ptolemy so closely, did likewise. Geocentric maps in the Ptolemaic tradition show the front sides of the constellations. As long as astronomers referred to stars by their position within a constellation—such as being located in the left arm of Boötes—the convention of geocentric fronts and external backs remained important. Yet many cartographers disregarded it. The maps in Bayer's *Uranometria* are geocentric while the figures are mixed (CAT. NOS. 2–5). Schiller's maps are external, but the figures are seen face on because, he explained, it would be an indignity to have saints and patriarchs show their backside (CAT. NOS. 70–73).

The ecliptic coordinate system defines celestial positions with reference to the ecliptic, the apparent annual path of the Sun. The equatorial system relates positions to the celestial equator, a great circle representing the extension of the equator of the Earth. Gradually, astronomers relinquished the ecliptic reference system in favor of the equatorial, and this shift is reflected on numerous maps. The Dürer planispheres, based on Ptolemy, extend from the ecliptic poles to the ecliptic. While still fundamentally ecliptic, the maps in Schiller's atlas carry coordinates of both systems.

The best celestial atlas at the end of the seventeenth century was the *Firmamentum Sobiescianum* by Johannes Hevelius, an astronomer with a private observatory in Danzig (CAT. NOS. 51–53). Like many scientists before and since, Hevelius understood the wisdom of flattering those who were or who might become his patrons. Accordingly, among the nine new constellations he placed in the sky was Scutum Sobiescianum, the shield of John Sobieski, King of Poland (CAT. NO. 53). Two of Hevelius's constellations relate directly to his observing practice. Although he had access to telescopes, he was a skillful enough observer to be wary of relying on them as they then were. Sextans Uraniae, a tiny constellation squeezed in between Leo and Hydra, represents one of the naked-eye instruments with which he had re-observed the northern stars (CAT. NO. 51). For the southern stars, however, Hevelius relied on the telescopic observations made by Edmund Halley, the famous English astronomer who, early in his career, spent 1677–78 observing on the island of St. Helena and visited Hevelius in 1679.

The first great celestial atlas based entirely on telescopic observations was the *Atlas Coelestis* of John Flamsteed. Although instrumental and theoretical advances soon made even better catalogues possible, Flamsteed's catalogue was notably superior to those that preceded it. Equipped not only with telescopes, but also with micrometers and improved clocks and theories of refraction, Flamsteed was able to reach a new level of accuracy as well as to record stars as faint as seventh magnitude. As an observer well aware of the cost and difficulty of obtaining fine instruments, Flamsteed appreciated the value of the celestial maps. In his words the maps were to be "the glory of the work, and next the catalogue, the usefullest part of it." To ensure this usefulness Flamsteed drew the maps on a modified conical projection on which, he believed, "the appearance to the naked eye is less distorted than by any projection I have yet seen." The maps are geocentric and so, following Ptolemaic convention, the figures are seen face on. Arguing that Bayer had distorted some of the Ptolemaic constellations, Flamsteed had them redrawn so as to correspond with Ptolemy's descriptions. With an ambitious program and limited resources, Flamsteed did not live to see his work reach the public. The catalogue was published, posthumously, in 1725, and the *Atlas Coelestis* in 1729. Four copies made of Flamsteed's constellations appear in *Celestial Images* (CAT. NOS. 47–49 & 74).

In 1801 Johann Elert Bode of Berlin issued his *Uranographia*, the last great atlas embellished with constellation figures. It included 17,240 stars of magnitudes 1–8, double stars, nebulae, and star clusters, all well graduated and organized into 99 constellations (CAT. NOS. 11–14). Some of the constellations were politically motivated, such as Cor Caroli (for Charles I of England), Robur Carolinum (for Charles II), and Sceptrum Brandenburgicum (for Frederic William III). And some, such as the fourteen fashioned after instruments of art and science and placed in the southern skies by N. L. de Lacaille, represented Enlightenment ideals. As if these were not sufficient, Bode added five more: Felis (the cat of the French astronomer Lalande), Globus Aerostaticus (in honor of the Montgolfier brothers), Lochium Funis (a nautical log line), Machina Electrica, and Officina Typographica (in honor of the 350th anniversary of the invention of movable type).

As Bode's atlas indicates, known celestial objects by 1800 had become so numerous they left little room for the intricacies of the constellation figures, and the constellations had become too numerous and contorted for convenience. Sensible astronomers called for reform. They reduced the number of constellations, rationalized their boundaries, and increasingly omitted the constellation figures—lessening the charm and beauty of astronomical charts, and creating functional tools of modern science.

THE FUNCTION OF ARTISTIC FORM IN THE STUDY OF THE STARS

SAMUEL Y. EDGERTON

ASTRONOMY IS THE SCIENCE with the longest human tradition. Awareness of sky patterns is perhaps the first intellectual achievement distinguishing hominids from the other primates roaming the primordial jungle. Evidence of such awareness dates back more than thirty thousand years; Alexander Marschak, in his seminal *Roots of Civilization*, has argued that the oldest symbols of language, the most ancient manmade marks on bone shards and cave walls (which are the rudest beginnings of writing and picture-making), were invented to keep track of the stars.[1] The changing but consistently repeated shapes of the phasing Moon, for example, stimulated primitive humans to make the original celestial chart, probably no more than a piece of antler horn carved with crude notches. From this simple tool, our hunter-gatherer ancestor was able to predict when to plant, when the caribou would be returning, and in general how to survive by anticipating the seasons.

How then did the "nonfunctional" imagery of the luxurious maps in *Celestial Images* metamorphose from the practical ruins of our primitive ancestors? Why did Western Europeans in particular continue to lavish Renaissance classical-style decor on star maps, even into the nineteenth century, thus confusing with subjective art what should have remained a more objective, scientific tool? From the purely scientific viewpoint of our "two-cultured" society, one might wonder whether the Copernican Revolution would not have happened sooner if astronomers had paid less attention to pretty pictures and more to no-nonsense mathematics. While acknowledging the aesthetic quality of these artistic works, the modern scientist might nonetheless doubt that the examples in this exhibition ever contributed anything to the ascent of man from superstitious fear to scientific understanding, from medieval astrology to modern astronomy.

I would like to argue that so-called subjective art has always contributed to the advance of science; it still does today, even in twenty-first-century astronomy. Historians are increasingly aware of the correlation between artistic style in any given society and that society's definition of truth. "Reality" may be a philosophical absolute, but historically and culturally it is always a relative idea. In order to learn then how each culture construes the facts of nature, we must first understand its forms of art. Far better than words, a picture can reveal what people at various times "saw." Human eyesight, of course, has remained physiologically constant (the pure optics of vision are exactly the same for us as they were for prehistoric cave men). Nonetheless the psychology of perception, what Wölfflin called "ways of seeing," has evolved differently under varying cultural pressures. By understanding these changes in "seeing" which are manifested in changes in artistic style, we can better understand developments in the history of science.

A good example is the stylistic preference during the Middle Ages in Western Europe for gold as the dominant pigment in the backgrounds of paintings. Gold, to medieval Christians, indicated more than just a rare and decorative luster; it symbolized the very brilliance of God's eternal realm in Heaven. When the devout of the eleventh and twelfth centuries "saw" the sky, they were not looking to determine the physical color of atmosphere but the metaphysical presence of the Almighty. Later, in the mid-thirteenth century, when we observe artists suddenly using blue in their pictures to indicate sky, we may suspect that something more than mere artistic whim was changing. At this very moment, even among the Franciscan monks whose mother church was the newly frescoed basilica at Assisi, a whole new concept of mundane reality was being debated, a concept which was to become the foundation for the scientific revolution three centuries hence. Painting the sky blue may seem obvious today (even as we realize that beyond the Earth's atmosphere the sky is in fact black), but in the scientific context of the thirteenth century, such a color choice had profound implications. It meant not only that attention was being shifted toward understanding of the optical effects of air, but also toward a new notion of *space*. Heretofore, space, if medieval philosophers thought of it at all, was conceived of as finite; the spatial realm of the world was distinctly different from the "space" of God. Now, in the thirteenth century, the idea began to dawn that space was something infinite and uniform and extending everywhere, *including* the domain of God in the furthest heavens. Thus when Giotto the artist painted blue skies in his 1306 frescoes for the Arena Chapel in Padua, also depicting angels flying just like birds in this airy space, he was inadvertently laying the groundwork for Galileo in the seventeenth century. Giotto, in effect, was saying with his art what the great astronomer later proved with his telescope, that angels and asteroids and whatever else is in heavenly space, react to the same optical laws and are of more or less the same substance as physical objects on Earth. Interestingly enough, the European Space Agency's 1986 probe to Halley's Comet was named "Giotto" to commemorate the painter's precocious depiction of the "reality" of a comet in another Arena Chapel fresco. Thus, this linkage between artist and physical scientist continued even into the twentieth century.

Before Giotto and artistic construction of three-dimensional space by the Assisi painters, pictures, and especially diagrams of the stars and planets, always gave the appearance of a strictly two-dimensional world. While medieval philosophers well knew that the Earth was round, they had difficulty "seeing" in their minds' eye the implication of its spherical form. They believed that the planets and the stars revolved about the Earth in circles, and so they represented the universe as a nest of flat concentric rings without considering that when such a system is matched with empirical observation, some of the revolving bodies must collide. It was not until Copernicus published his *De revolutionibus orbium coelestium* ("On the Revolutions of the Heavenly Spheres") in 1543 that mathematical proof was offered showing that only when the Sun was at the center of the universe could the stars and planets revolve properly. Copernicus studied astronomy in Renaissance Italy when artists were redefining pictorial space in such a way that viewers could imagine the world in all its three-dimensionality. This same new perceptual attitude was adopted later by Copernicus, enabling him to "see" the sky no longer as flat and finite but as an isotropic continuum. The planets and stars could then be understood as moving parts of a three-dimensional machine. Not coincidentally then, Copernicus's conception of rotating heavenly bodies is like Leonard da Vinci's nearly contemporary designs for complex machinery, also carefully imaged so that the moving parts did not bump into one another.

We take it for granted today that the sky is open space extending to infinity and filled with myriad, moving volumes. Yet such a notion was not so easy to grasp until the advent of the Renaissance. People needed "visual aids" to jar their comprehension. The application of what appears at first glance to be purely extraneous decor in the star maps here on display, like the draperies depicted as hovering incongruously in the sky in the illustrations of Mallet's 1683 *Eclipse* and *Phases of the Moon* (CAT. NOS. 60 & 61), actually performed a psychological function. They helped the observer to imagine the celestial phenomena described in mathematical notations as really existing in three dimensions. In other words, the Renaissance-style pictures, even of mythological beasts and other outlandish figures decorating these charts, were intended not just to embellish, but also to stimulate the viewer's mental image of the stars as in-the-round volumes in infinitely extending space.

The French word "renaissance" refers to the rebirth of art forms first conceived by the ancient Greeks and

Romans. Artists of the fifteenth and sixteenth centuries finally rediscovered what the ancients already knew but which had been forgotten during the Middle Ages: namely, linear perspective, light-and-shadow (*chiaroscuro*) rendering and correct anatomical drawing of the human figure. The Greeks, in fact, were so fascinated by figure drawing that they imagined physiognomical forms everywhere including in the willy-nilly arrangement of stars in the sky. Their system of cataloguing the stars by grouping them in anthropomorphic shapes was a brilliant art of descriptive science. No matter how far-fetched the associations, such identification of the constellations with familiar early personifications provided reassurance that man could know the heavens. Thus, when Renaissance artists and scientists picked up where the ancients left off, they found to their good fortune a ready-made system for representing the stars and planets in familiar, tangible pictorial forms.

Some three centuries after Giotto and the Assisi painters inaugurated the Renaissance, another Florentine, Galileo, looked through a telescope and, framing the Moon just as Giotto windowed his painted angels, realized for the first time that the lunar surface was covered with voluminous mountains and craters. Besides being a great scientist, Galileo was also an expert draughtsman, which should not be surprising considering his residence in a city of such artistic fame as Florence. He had learned perspective and chiaroscuro so well that he drew his own illustrations for the book that announced his lunar discoveries in 1610.[2] Unquestionably, also, Galileo's Renaissance art training made it easier for him to "see" the true nature of the Moon's landscape. He had in fact studied from treatises like Daniel Barbaro's *La practica della perspettiva* (Venice, 1568), which required him to master such problems as the drawing of shaded spheres with raised protuberances. Galileo was quite explicit in his description of the Moon that the dark spots on its surface were caused not by any mysterious "vapour" (as some medieval philosophers believed) but by cast shadows, exactly as he must have often drawn himself while practicing perspective. It is also noteworthy that some six months before Galileo, an English astronomer, Thomas Harriot, had similarly looked through a telescope at the lunar surface. Harriot, who had no practical artistic experience and whose native land was still resisting the Italian Renaissance, had no frame of reference with which to compare what he saw when observing the magnified Moon. Consequently, that "strange spottednesse" (as he called it) remained inexplicable.[3]

It was no accident, I believe, that the scientific revolution of the seventeenth century followed immediately upon the artistic revolution of the Italian Renaissance. Not only was astronomy inspired by art but so also were many other of the evolving physical and biological sciences. One might even say that the spread of the scientific revolution in Europe was directly proportional to the speed with which the various countries on the continent adapted to Renaissance style. By the time Sir Isaac Newton wrote the laws that were to be operative for all physical science including astronomy until the twentieth century, English art was at last succumbing to Italian Renaissance domination. In fact, the basic assumption of Italian Renaissance art as first established by Florentine painters in the fourteenth and fifteenth centuries was that the pictorial world would be organized so that all depicted objects appear to be seen from one eye-point in a single moment of time. This is exactly in keeping with Newton's law of inertia, which explains how all physical bodies in the real world, including the stars, rest or move in relation to a fixed reference point on Earth. Renaissance art in effect "depicts" Newton's physics. Indeed, all the decorated celestial charts in the present exhibition belong to what, since Thomas Kuhn, has been called the "Newtonian paradigm": images of stars and planets artificially stopped in space and time, as if observed from a single and immoveable viewpoint, just as Giotto imagined his angels.

Having noted all this about "Newtonian" celestial maps in the present exhibition, it is fascinating to consider that in the very decade that the most recent star maps here displayed were published (Burritt's *Atlas*, 1835, CAT. NO. 15), two men were born whose life work would lead to the destruction of the "Newtonian paradigm" forever: James Clerk Maxwell (born in 1831) and Paul Cézanne (born in 1839), the former a British physicist, the latter a French painter. Certainly, neither ever heard of the other, and even if they had they would not have appreciated any connection between their respective pursuits; yet each in his own way discovered shortcomings in the heretofore unassailable Renaissance way of imaging the world. It was Maxwell the scientist who, while studying electricity and electromagnetism, came to the conclusion that these phenomena could not be explained by conventional, Newtonian physics; that is, unless real space be filled with a radically unorthodox, hypothetical substance called "ether." At the same time, the painter Cézanne arrived in Paris to propound his theory concerning pictorial space, henceforth no longer subject to the traditional laws of Renaissance linear perspective.

It would be incorrect for me to push the relationship between Maxwell's science and Cézanne's art too far, for neither really broke with the prevailing system. Maxwell hoped his postulated "ether" would explain the strange behavior of electricity and light *a la* standard Newtonian physics. Cézanne, in turn, never repudiated completely the old Giotto-style window notion of a picture. Nonetheless, these two contemporary geniuses managed to sow the seeds of the most profound crises in the history of science and art. As everyone knows, it was their respective followers, Albert Einstein and Pablo Picasso, who actually produced the revolution of Modernism. Again by coincidence, in the same year, 1905, when Einstein published his special theory of relativity, Picasso had just visited the retrospective exhibition of Cézanne's paintings and came to a decisive conclusion of his own, restructuring art in a radical new style to be called Cubism.

Once more, the direct links between Einstein's science and Picasso's paintings are tenuous, yet it is a fact that both finally broke irrevocably with the old, classical way of looking at the universe. Einstein, for instance, rejected "ether." There was no need to invent it in order to retain Newton's law; the law itself was wrong and must be replaced by a new concept of "reality." At the same time, Picasso rejected the old "picture-as-window" concept that forced all depicted objects on the canvas to be ordered from a single eyepoint. Like Einstein, he took into consideration that the observer is never stationary but constantly moving relative to the objects observed. Both Einstein and Picasso were able to liberate their imaginations from the tyranny of linear perspective, and thus "saw" nature in ways no one had ever dreamed of before.

Einstein's relativity introduced astronomy to concepts impossible to visualize by any traditional means of imaging. How, for example, was one to depict in the mind's eye, let alone on paper, anti-matter black holes or endlessly exploding quasars in extra-galactic infinity? Interestingly, the conventions of modern, post-Cubist art, especially its free play with abstract color forms, have been appropriated by contemporary astronomers who find them remarkably useful for recording just such mind-boggling celestial phenomena. I am referring specifically to the "Image Processing Facilities" that have been installed in various Astronomy Departments around the world, where scientists, detecting new celestial presences with telescopes often operating at "invisible wavelengths," are able to convert these into beautiful color formations on television screens. This is done by means of a special silicon chip known as a CCD or "charged-coupled device," which translates light sensations from the distant nebulae into a matrix of digital pulses. Bright colors are then induced into these "pixels," as each sensitized point of the electronic grid is called. Each color represents a variation of the astral form's light and/or heat intensity.[4] Since these colors are artificial, they can only symbolize, just as the conventions of painting do, the phenomena of visual reality. The modern astronomer thus becomes as artist once again, although exchanging Renaissance perspective for the nonillusionary flat shapes of abstract art.

Perhaps the most exciting astronomical phenomenon which CCD technology has so far succeeded in depicting is that of "gravitational light deflection," the bending of light rays from far-distant galaxies as they pass close by another massive heavenly body. Einstein predicted this should happen. Today the CCD monitor can record it in the form of a colorful design of double images. Scientists call this process "gravitational lensing," but it also looks remarkably like the sort of painting Wassily Kandinsky was doing in 1914. In other words, what Einstein early in the century could conceive of mathematically has at last been scientifically represented in those very pictorial shapes which at the time were regarded as "unrealistic." The celestial map of modern astronomy now depends on new modes of abstract art, just as in earlier times the stars were made understandable by means of Renaissance painting conventions. Who can say when another astronomer, educated in the imagery of the CCD screen, will "see" something in the heavens never noticed before, and make a startling discovery just as Galileo did when he applied his perspective-trained eyesight to the surface of the Moon? If and when this happens, it will be due in no small measure to the heritage and tradition of the pictures here exhibited in *Celestial Images*.

1. Alexander Marshack, *The Roots of Civilization* (New York: McGraw-Hill, 1972).

2. Galileo Gallilei, *Sidereus nuncius* (Venice, 1610).

3. See my "Galileo, *disegno*, and the 'strange spottednesse' of the Moon," *Art Journal*, Fall 1984.

4. See James Cornell, "Video Stars: Image Processing for Astronomy," *Smithsonian Institution Research Reports*, Winter, 1983, p. 2.

PLATES

CAT. NO. 1 Frans Afferden, *Plani-Spherium Coeleste* from *El Atlas abreviado*, 1696

CAT. NO. 2 Johann Bayer, *Cassiopeia* from *Uranometria*, 1639

CAT. NO. 3 Johann Bayer, *Cygnus* from *Uranometria*, 1639

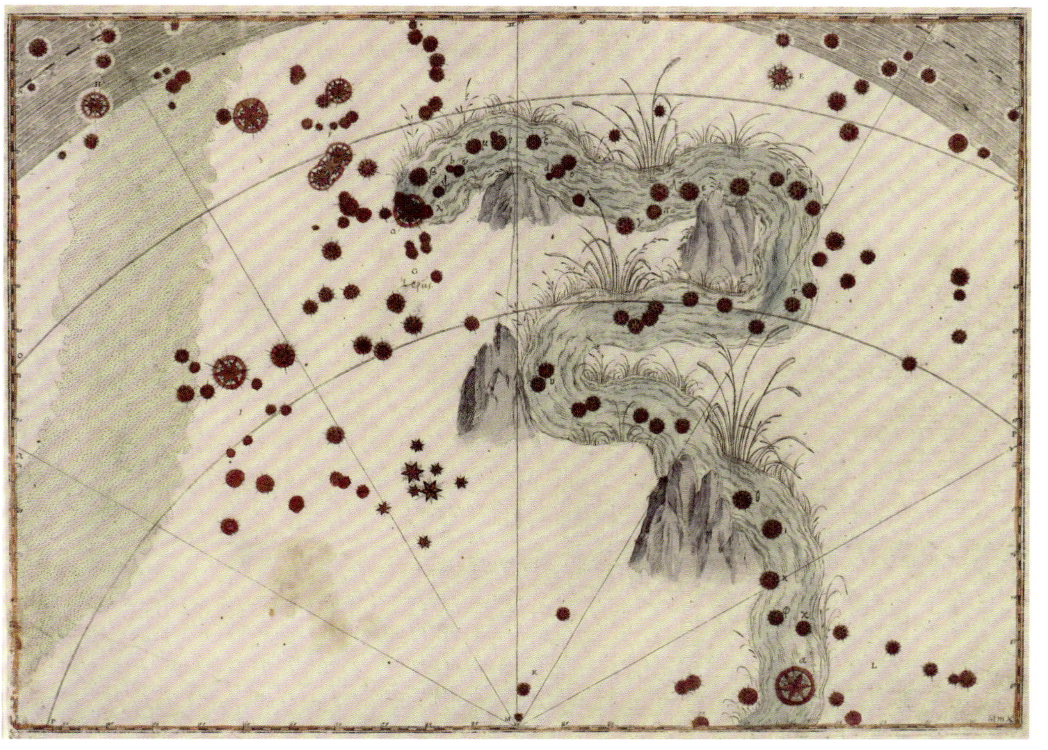

CAT. NO. 4 Johann Bayer, *Eridanus* from *Uranometria*, 1639

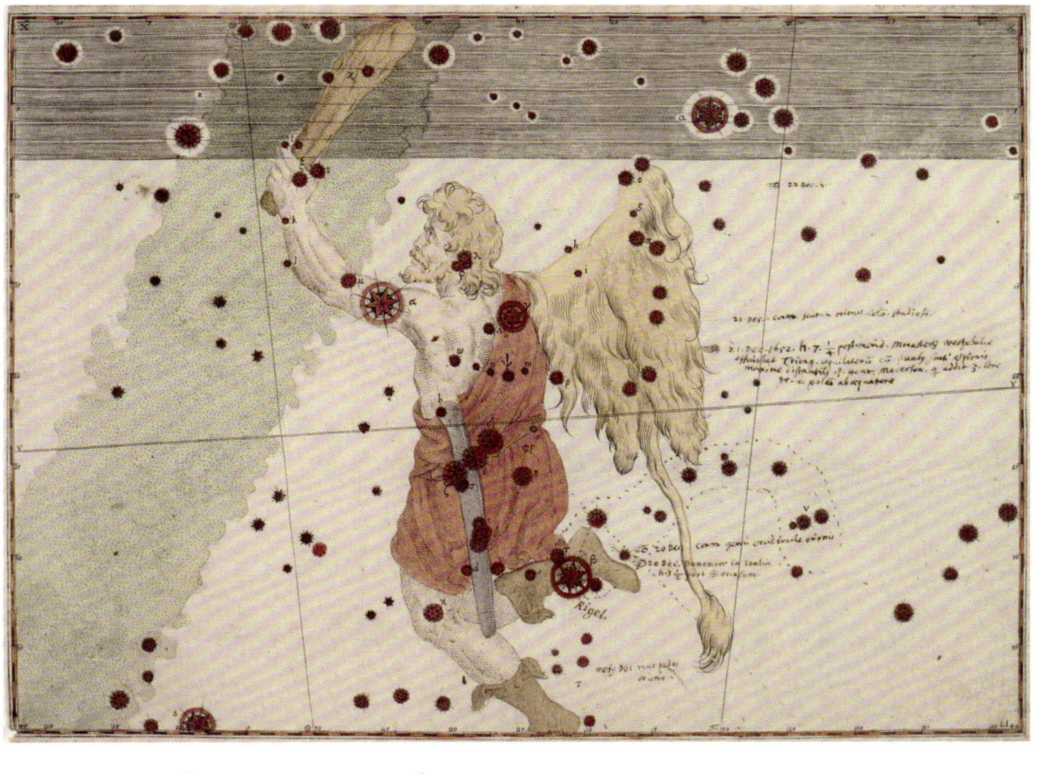

CAT. NO. 5 Johann Bayer, *Orion* from *Uranometria*, 1639

CAT. NO. 6 John Bevis, *Corona Australis* from *Atlas Celeste*, 1750

CAT. NO. 7 John Bevis, *Crator and Corvus* from *Atlas Celeste*, 1750

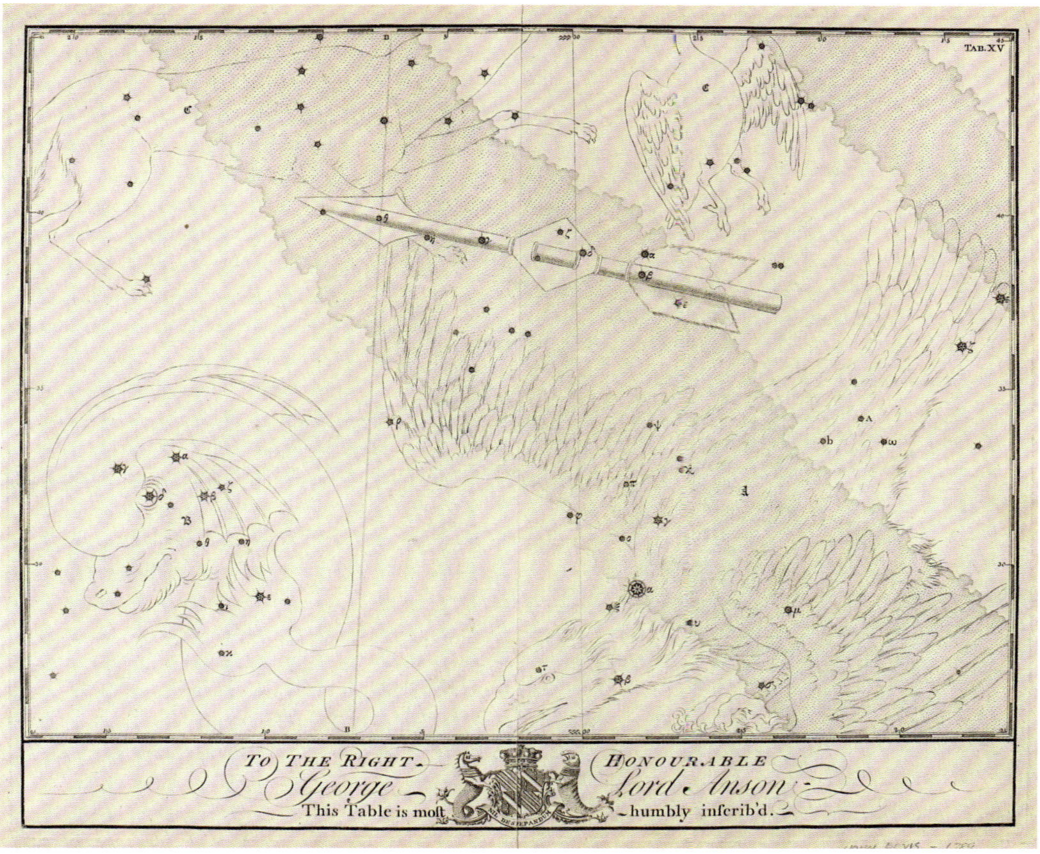

CAT. NO. 8 John Bevis, *Sagitta* from *Atlas Celeste*, 1786

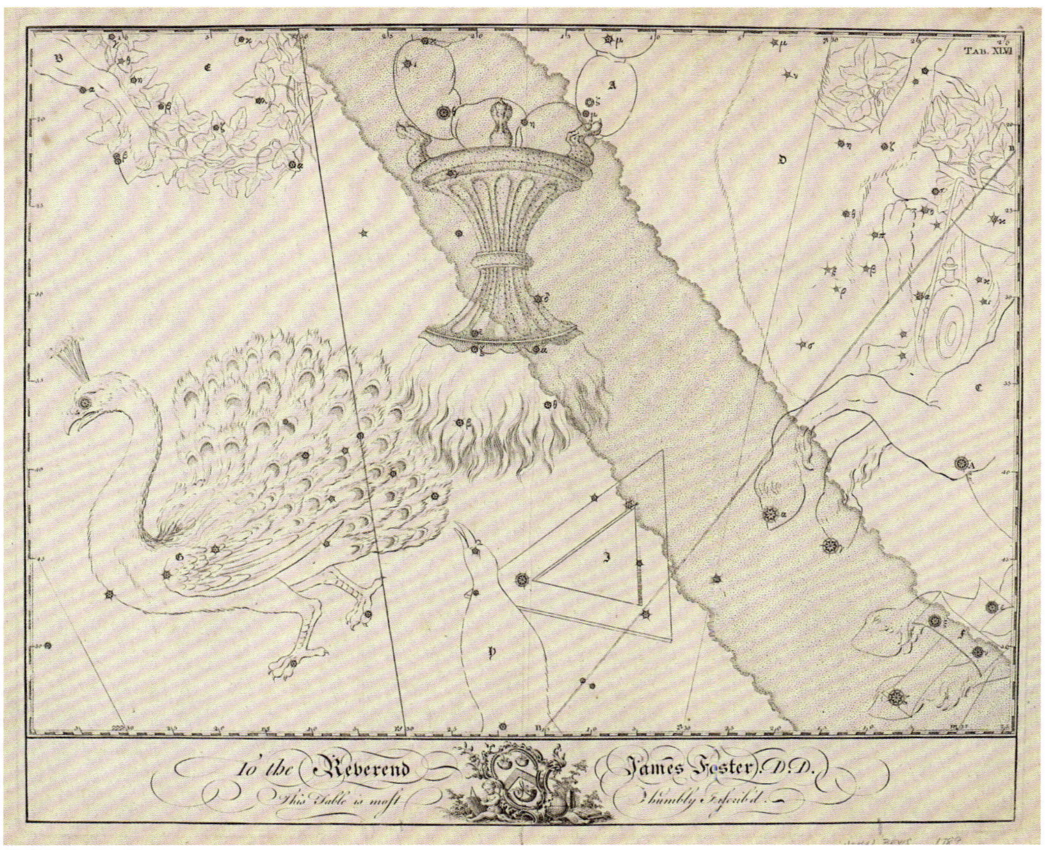

CAT. NO. 9 John Bevis, *Ara* from *Atlas Celeste*, 1786

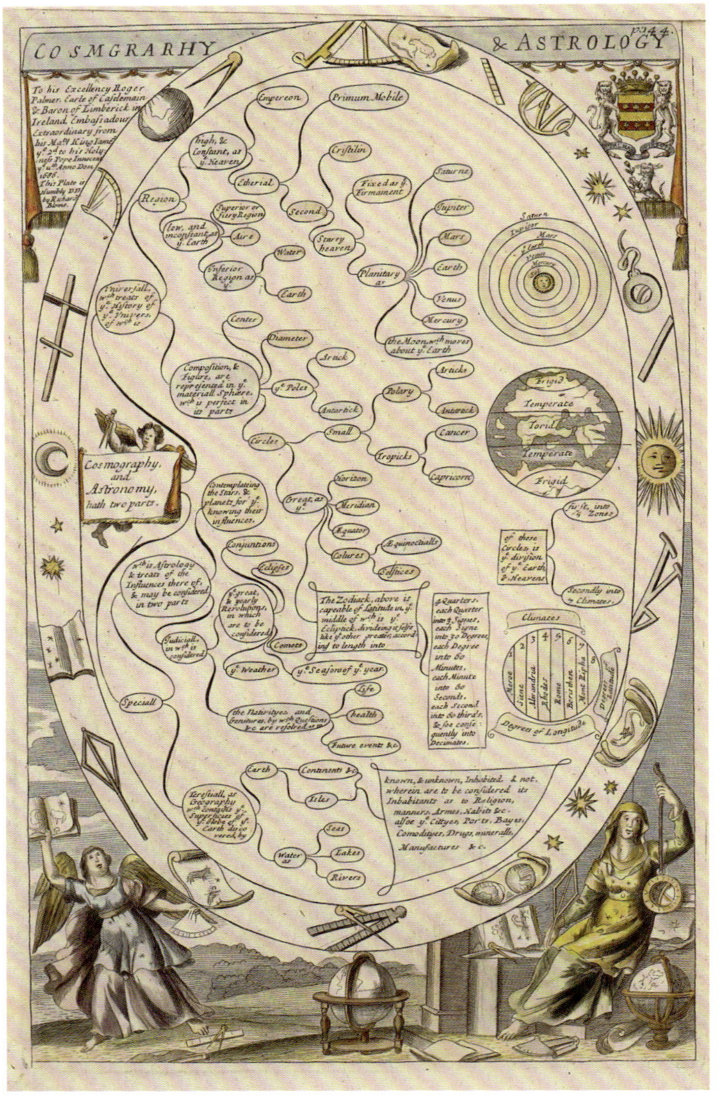

CAT. NO. 10 Richard Bloom, *Cosmography and Astrology* from *The Gentleman's Recreation*, c. 1686

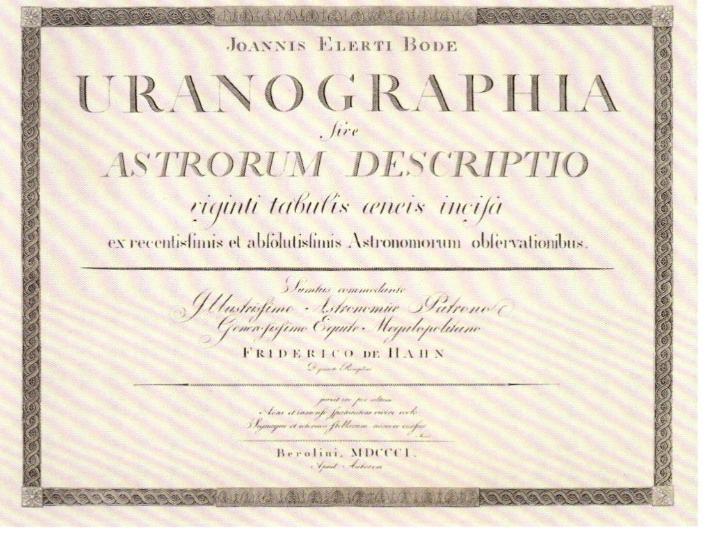
JOANNIS ELERTI BODE

URANOGRAPHIA

sive

ASTRORUM DESCRIPTIO

viginti tabulis aeneis incisa

ex recentissimis et absolutissimis Astronomorum observationibus.

FRIDERICO DE HAHN

Berolini, MDCCCI.

CAT. NO. 11 Johann Elert Bode, *Title page* from *Uranographia sive Astrorum Descriptio vinginti tabulis aeneis incisa ex recentissimis et absolutissimis Astronomorum observationibus*, 1801

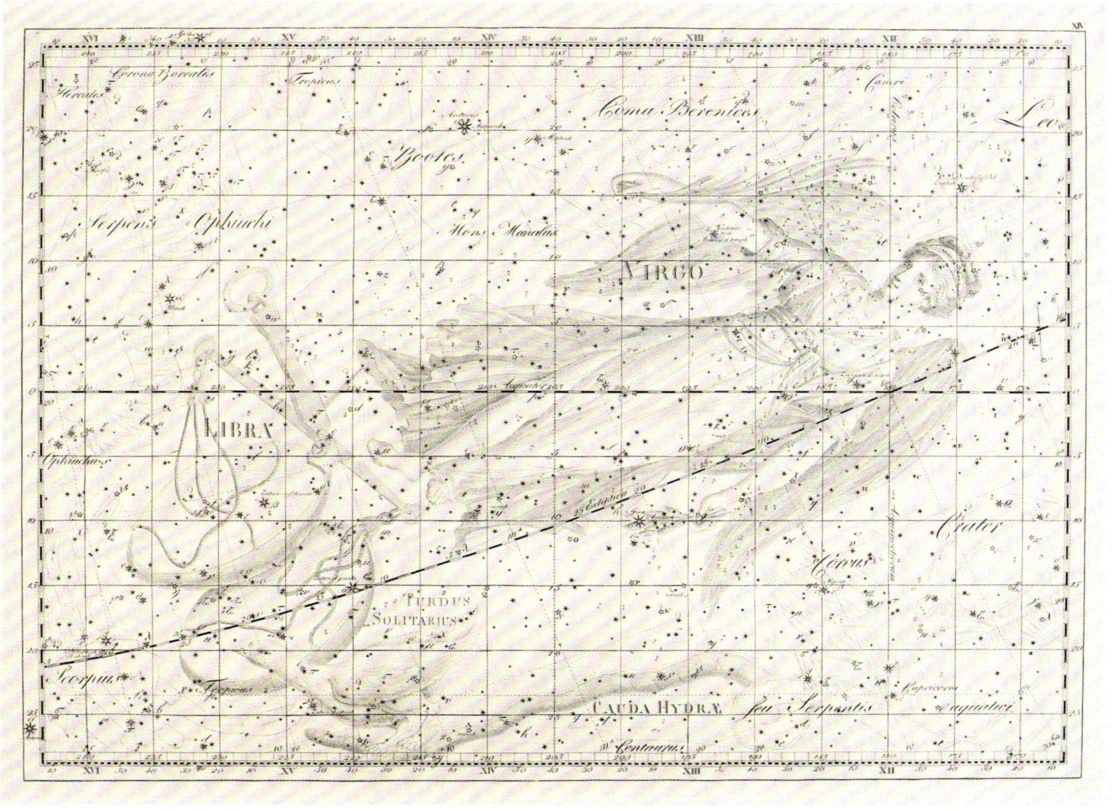

CAT. NO. 12 Johann Elert Bode, *Virgo and Libra* from *Uranographia sive Astrorum Descriptio vinginti tabulis aeneis incisa ex recentissimis et absolutissimis Astronomorum observationibus*, 1801

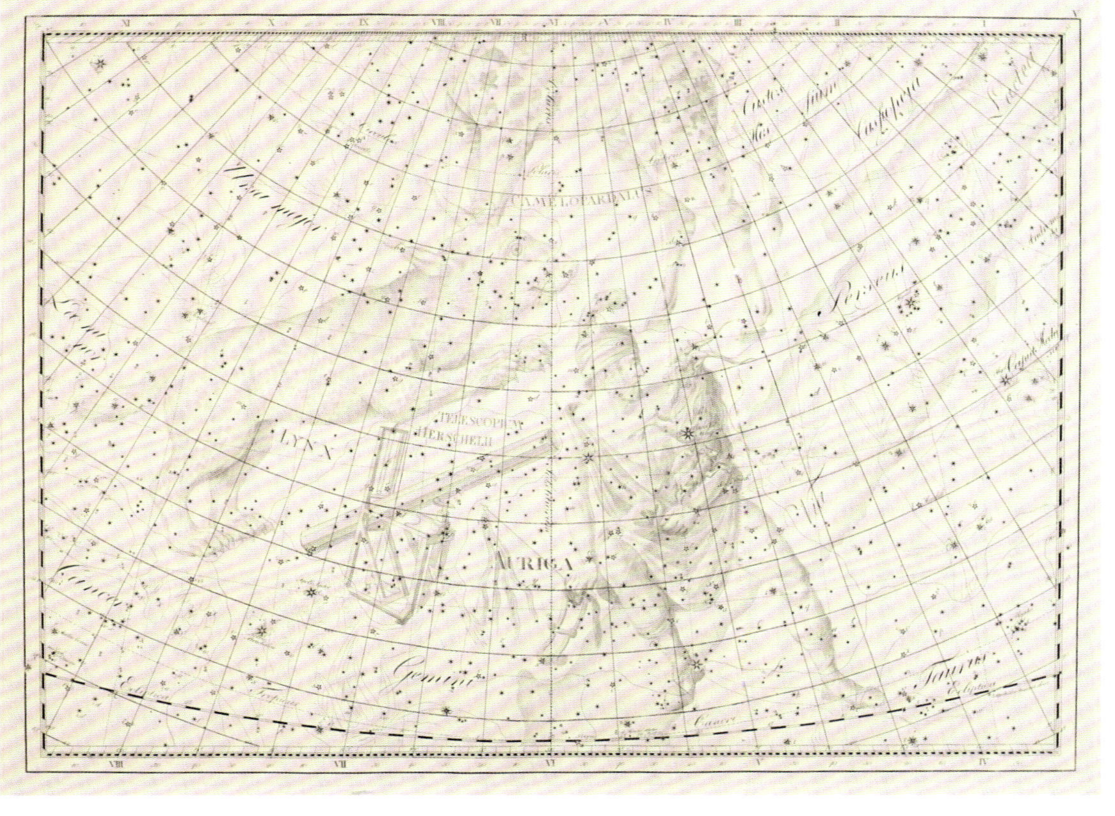

CAT. NO. 13 Johann Elert Bode, *Auriga and Lynx* from *Uranographia sive Astrorum Descriptio vinginti tabulis aeneis incisa ex recentissimis et absolutissimis Astronomorum observationibus*, 1801

CAT. NO. 14 Johann Elert Bode, *Southern Pole Constellations* from *Uranographia sive Astrorum Descriptio vinginti tabulis aeneis incisa ex recentissimis et absolutissimis Astronomorum observationibus*, 1801

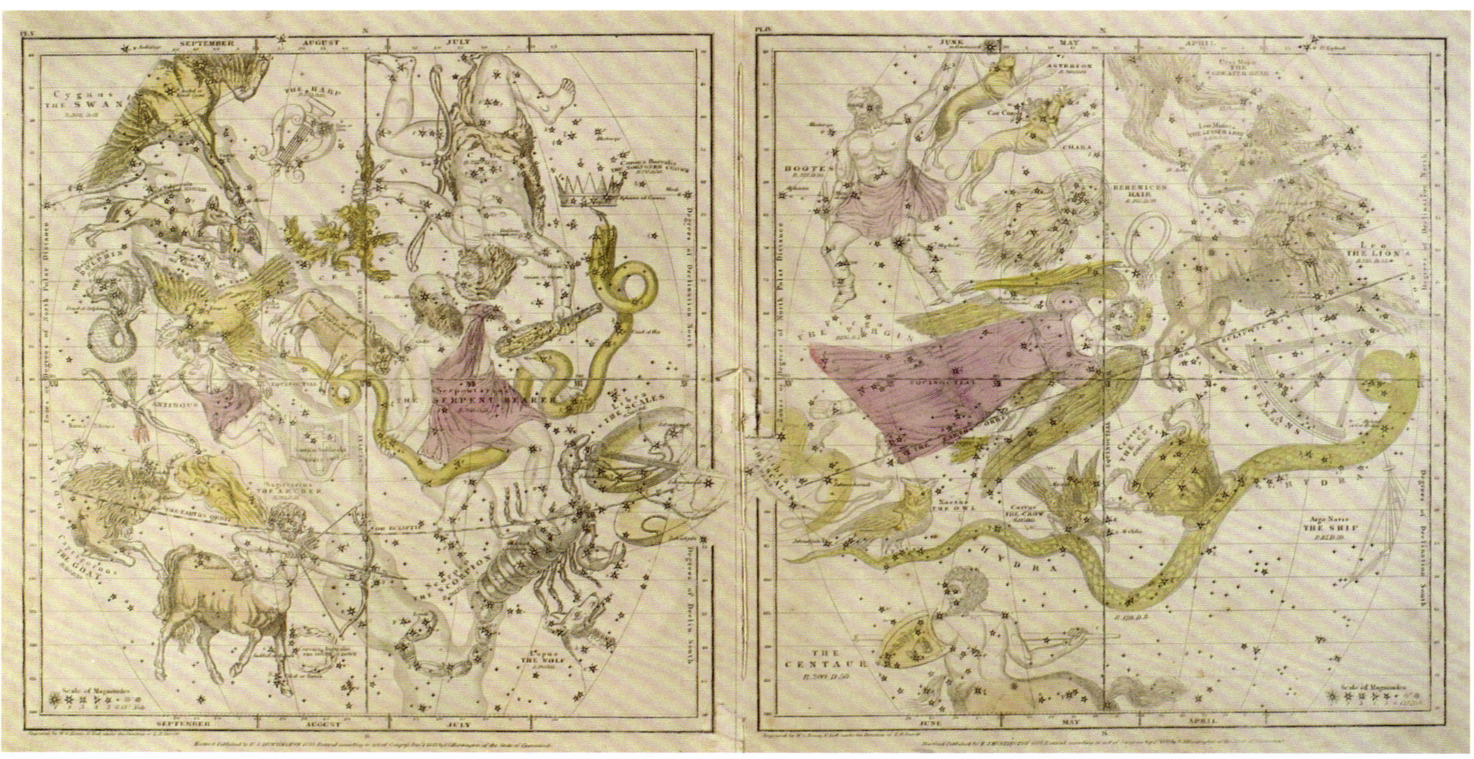

CAT. NO. 15 Elijah Hinsdale Burritt, *The Visible Heavens in July, August and September*, and *The Visible Heavens in April, May and June* from *The Atlas Designed to Illustrate the Geography of the Heavens*, 1835

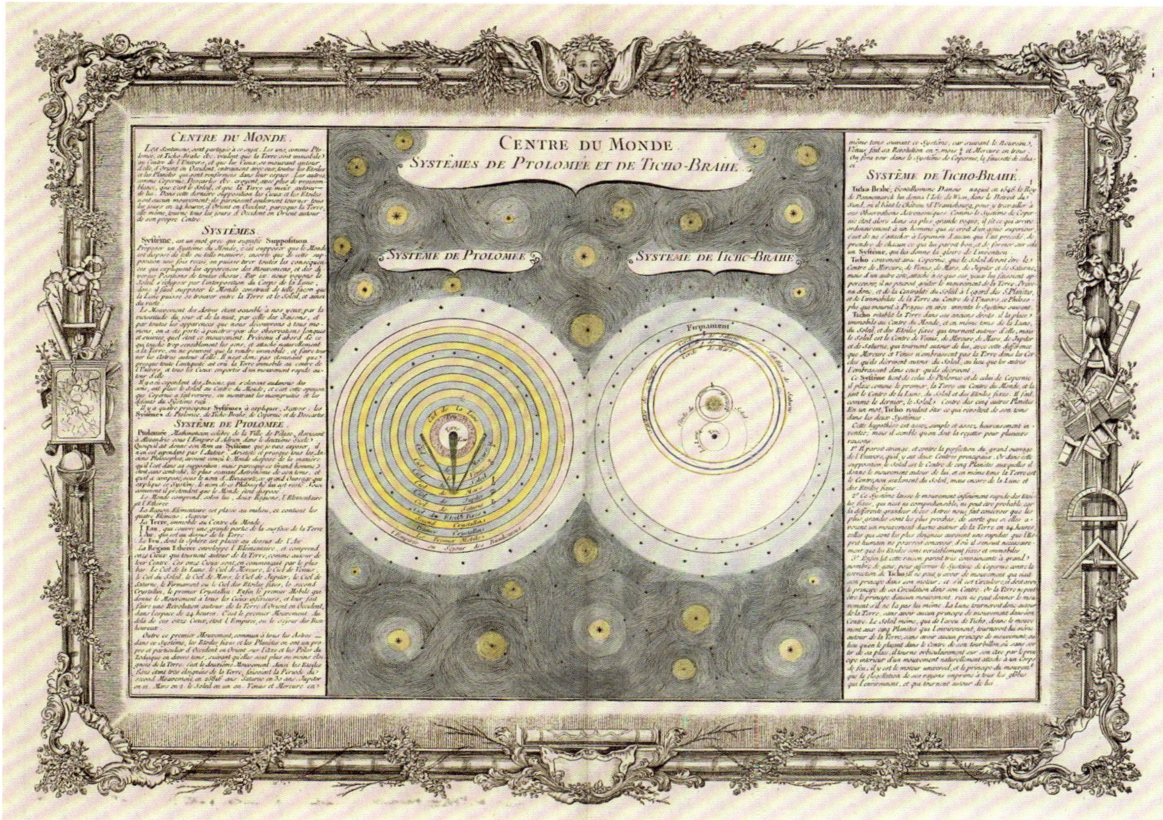

CAT. NO. 16 Claude Buy de Mornas, *Systèmes de Ptolomée et de Ticho Brahé* from *Atlas Méthodique et Elémentaire de Géographie et d'Historie*, 1761

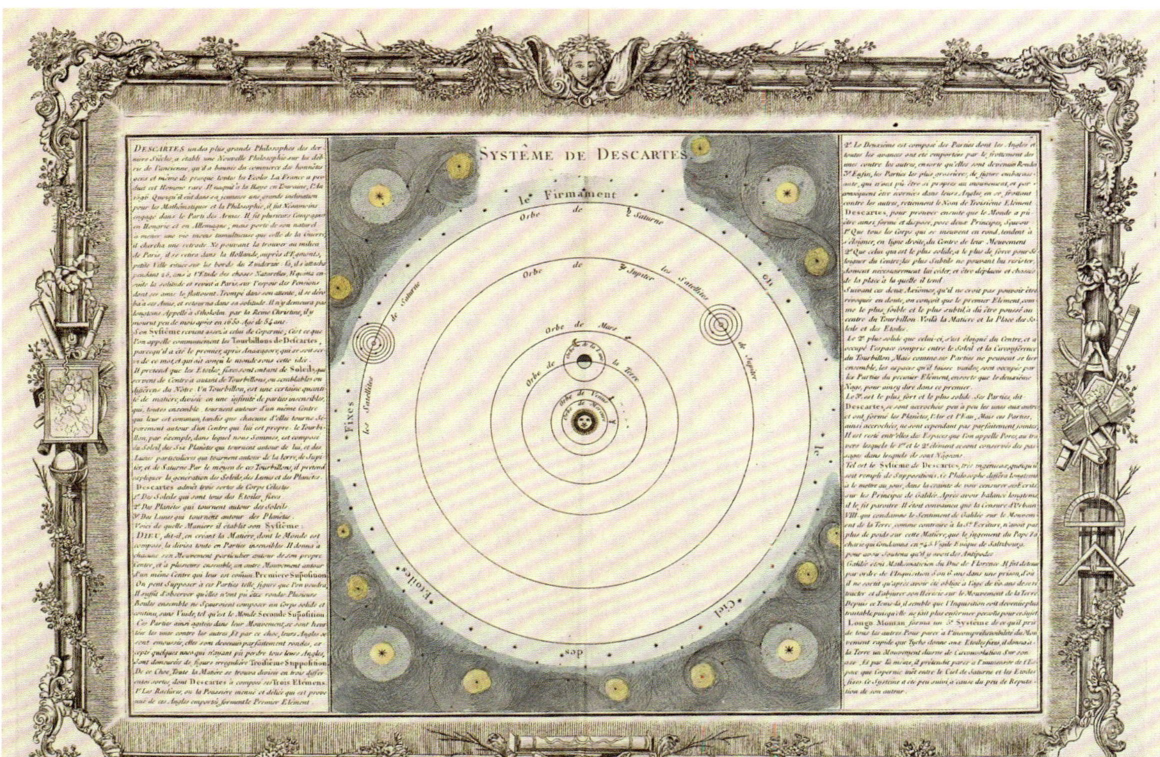

CAT. NO. 17 Claude Buy de Mornas, *Système de Descartes* from *Atlas Méthodique et Elémentaire de Géographie et d'Historie*, 1761

CAT. NO. 18 Giovanni Maria Cassini, *Celestial Globe Gores: Cassiopeia, Perseo, Orione* from *Globo Celeste*, 1792

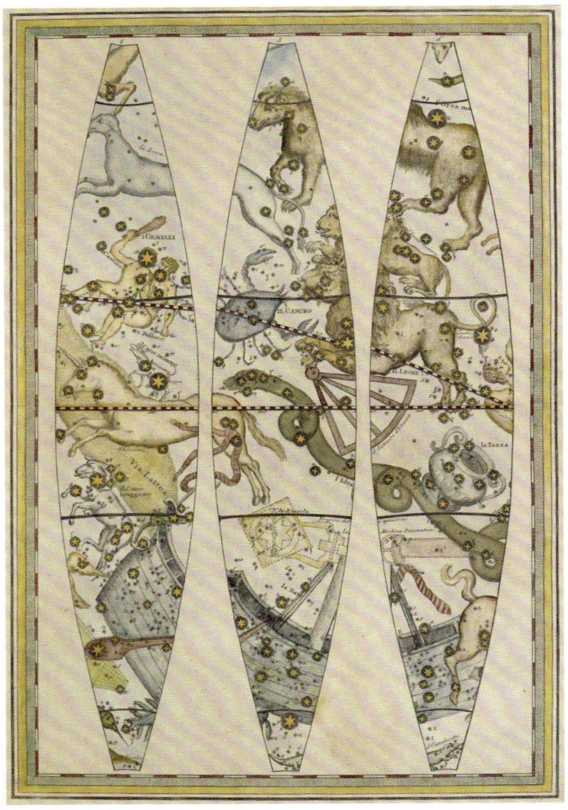

CAT. NO. 19 Giovanni Maria Cassini, *Celestial Globe Gores: Gemelli, Cancro, Leone* from *Globo Celeste*, 1792

CAT. NO. 20 Giovanni Maria Cassini, *Celestial Globe Gores: Vergine, Boote, Ercole* from *Globo Celeste*, 1792

CAT. NO. 21 Giovanni Maria Cassini, *Celestial Globe Gores: Sagitarrio, Capricorne, Aquario* from *Globo Celeste*, 1792

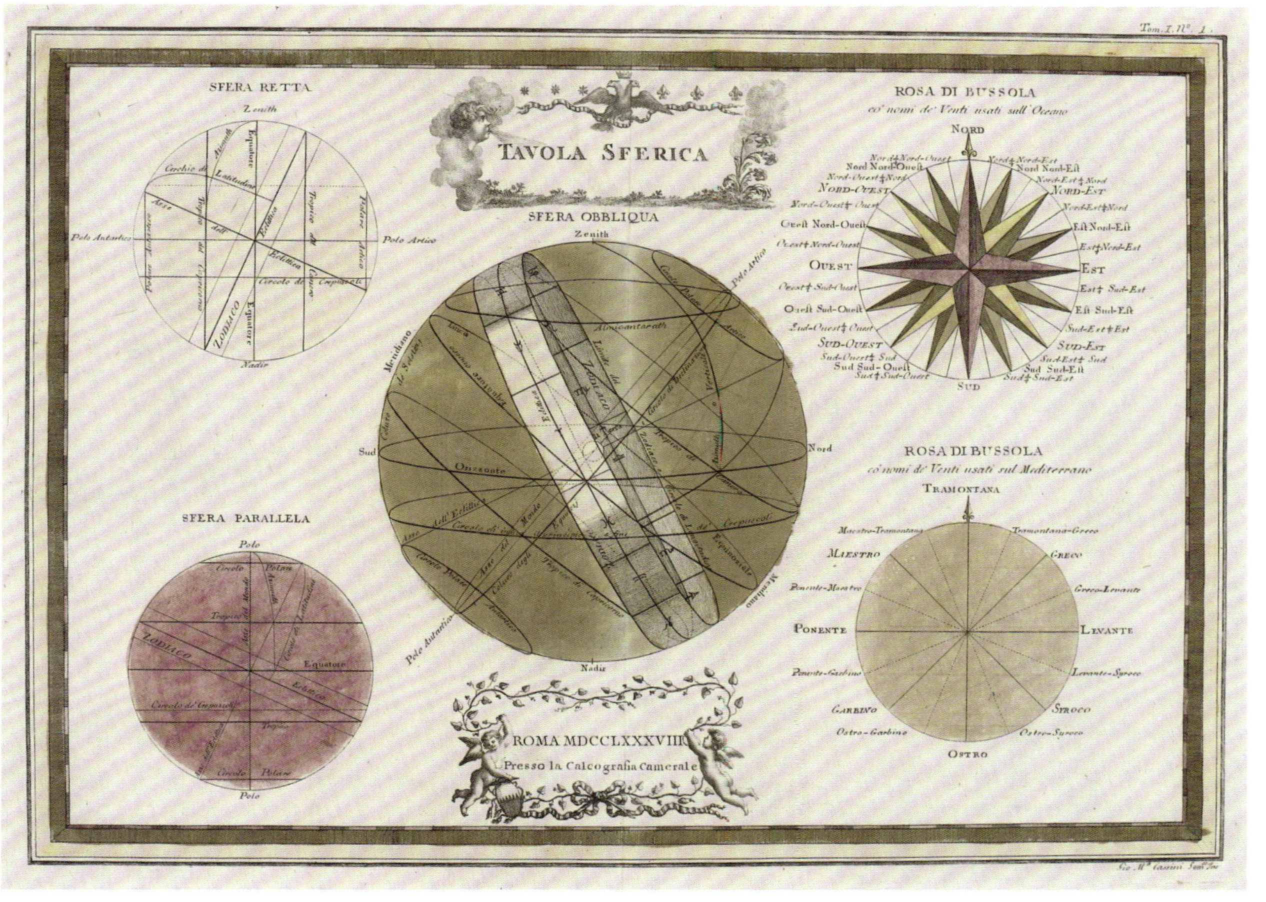

CAT. NO. 22 Giovanni Maria Cassini, *Tavola Sferica*, 1788

CAT. NO. 23 Andreas Cellarius, *Situs Terrae Circulus Coelestibus* from *Atlas Coelestis seu Harmonia Macrocosmica* (1st edition, 2nd printing), 1662

CAT. NO. 24 Andreas Cellarius, *Typus Selenographicus* from *Atlas Coelestis seu Harmonia Macrocosmica* (1st edition, 2nd printing), 1662

CAT. NO. 25 Andreas Cellarius, *Planisphaerium Braheum* from *Atlas Coelestis seu Harmonia Macrocosmica* (1st edition, 2nd printing), 1662

CAT. NO. 26 Andreas Cellarius, *Planisphaerium Braheum* from *Atlas Coelestis seu Harmonia Macrocosmica* (2nd edition), 1708

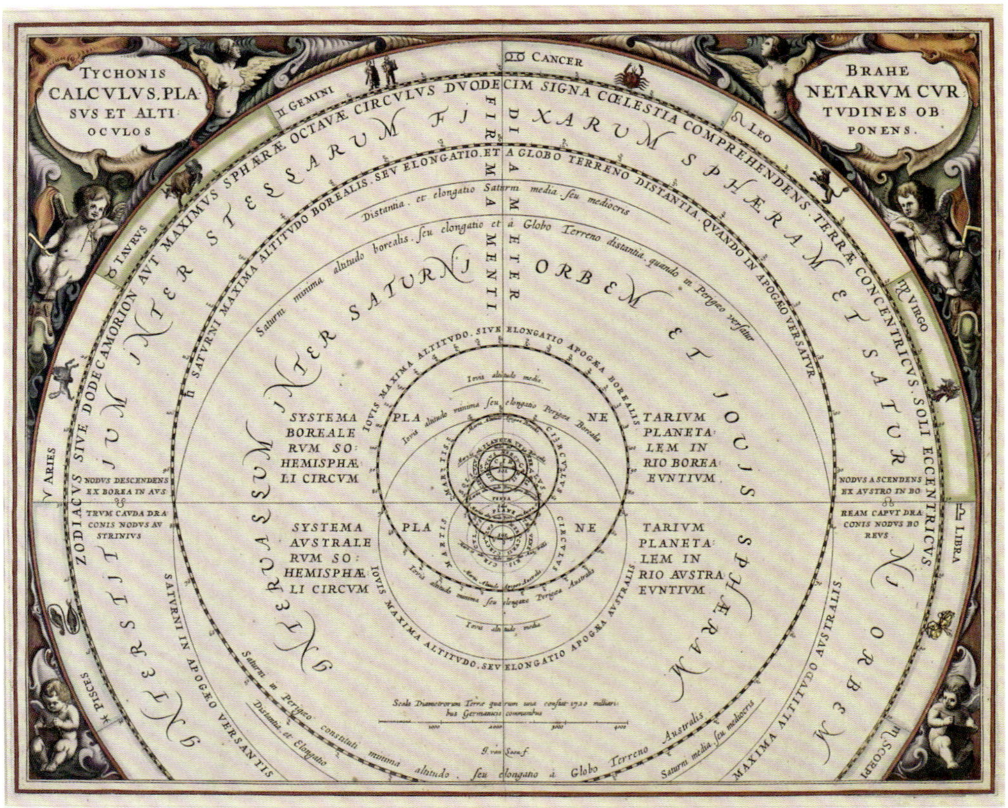

CAT. NO. 27 Andreas Cellarius, *Tychonis Brahe Calculus Planetarium* from *Atlas Coelestis seu Harmonia Macrocosmica* (1st edition, 1st printing), 1660

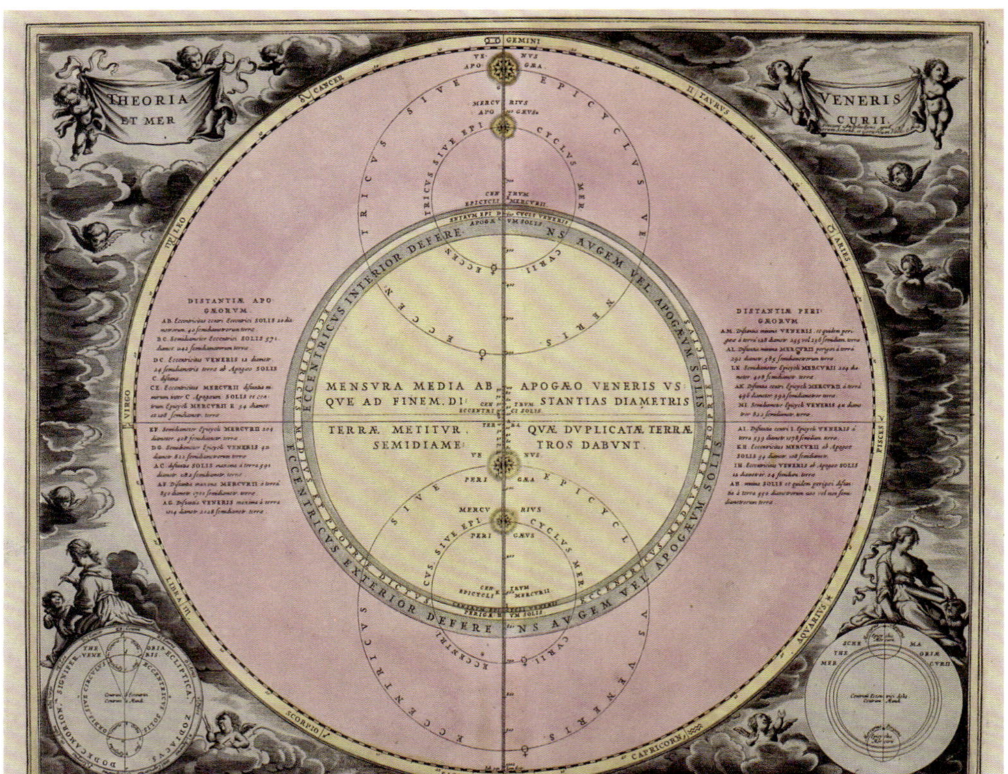

CAT. NO. 28 Andreas Cellarius, *Theoria de Veneris et Mercurii* from *Atlas Coelestis seu Harmonia Macrocosmica* (2nd edition), 1708

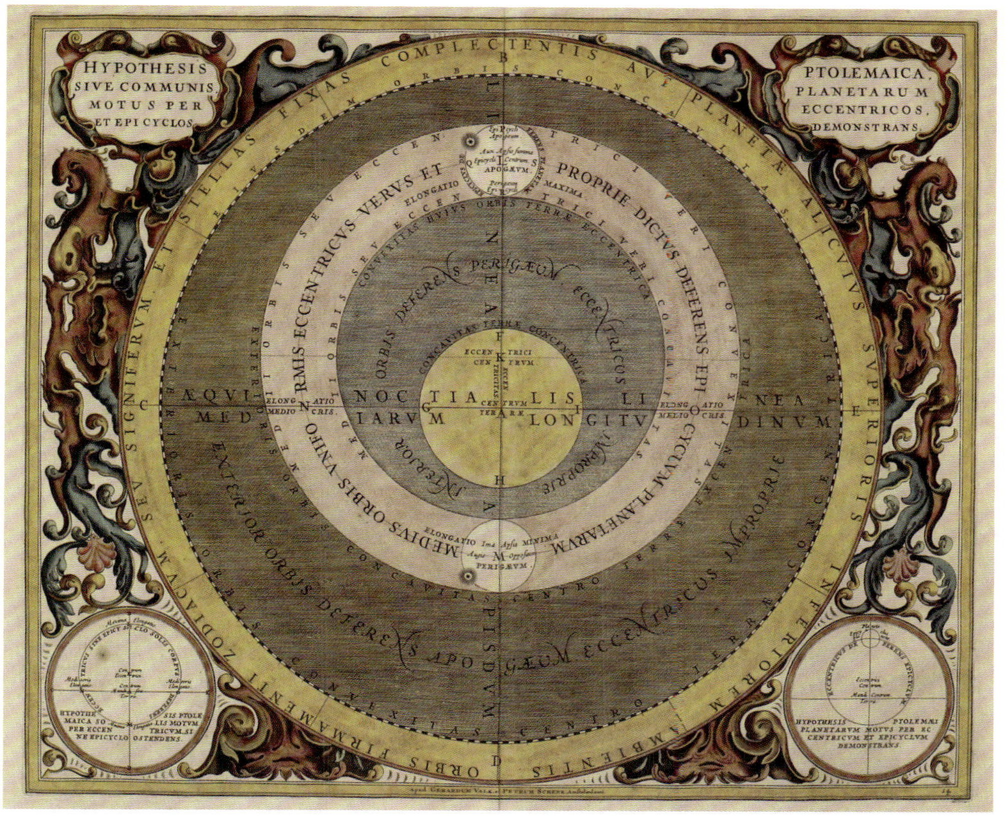

CAT. NO. 29 Andreas Cellarius, *Hypothesis Ptolemaica* from *Atlas Coelestis seu Harmonia Macrocosmica* (2nd edition), 1708

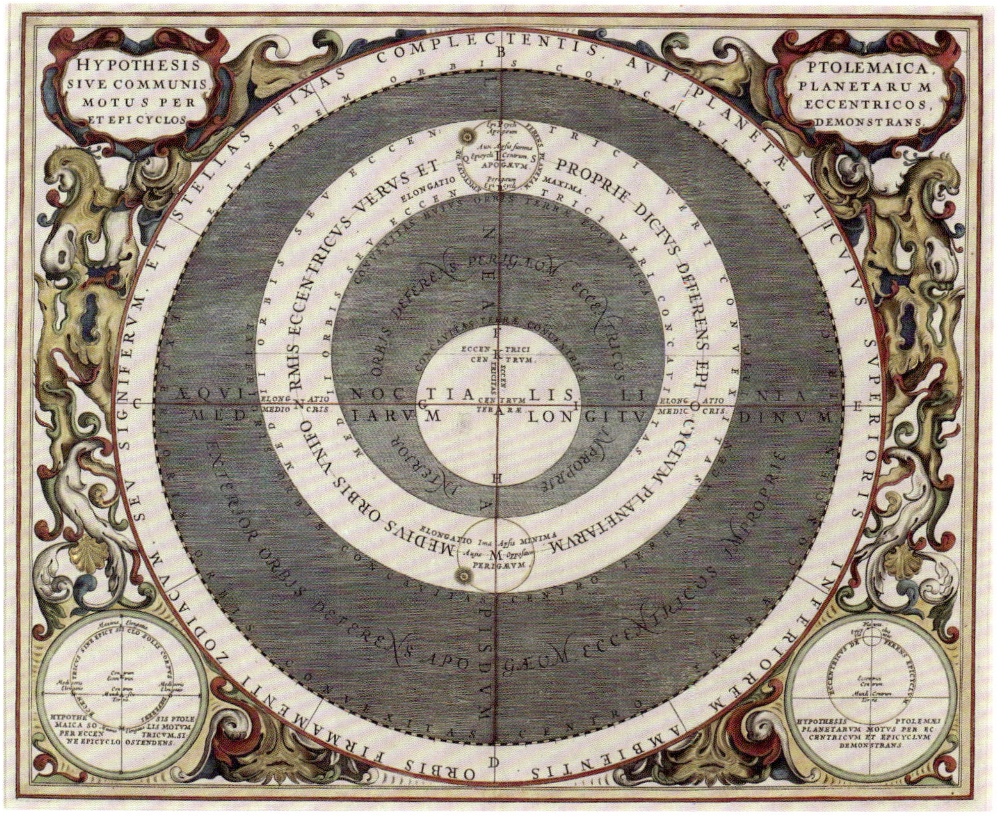

CAT. NO. 30 Andreas Cellarius, *Hypothesis Ptolemaica* from *Atlas Coelestis seu Harmonia Macrocosmica* (1st edition, 2nd printing), 1662

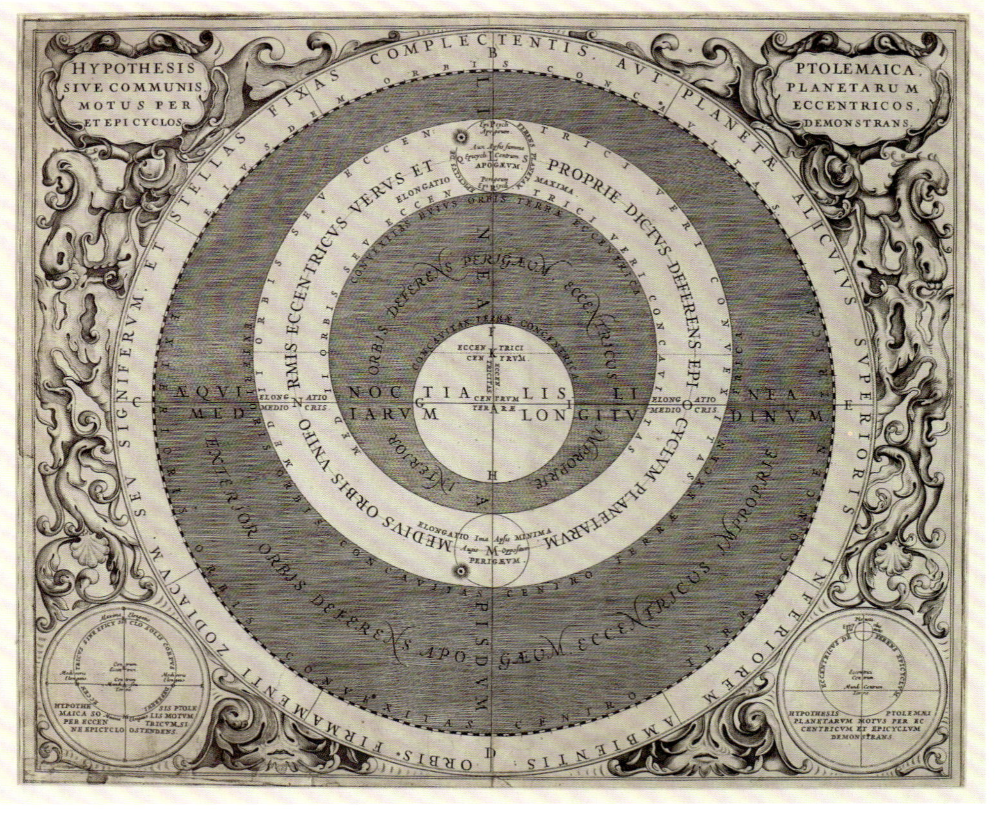

CAT. NO. 31 Andreas Cellarius, *Hypothesis Ptolemaica* from *Atlas Coelestis seu Harmonia Macrocosmica* (1st edition, 1st printing), 1660

CATALOGUE IMAGES TO THE RIGHT

CAT. NO. 32 Vincenzo Maria Coronelli, *Hercules (Celestial Globe Gore)* from *Oribis Coelestis Typus*, twentieth-century re-strike from original plates dated 1693

CAT. NO. 33 Vincenzo Maria Coronelli, *Serpens (Celestial Globe Gore)* from *Oribis Coelestis Typus*, twentieth-century re-strike from original plates dated 1693

CAT. NO. 34 Vincenzo Maria Coronelli, *Gemini (Celestial Globe Gore)* from *Oribis Coelestis Typus*, twentieth-century re-strike from original plates dated 1693

CAT. NO. 35 Vincenzo Maria Coronelli, *Canis Major (Celestial Globe Gore)* from *Oribis Coelestis Typus*, twentieth-century re-strike from original plates dated 1693

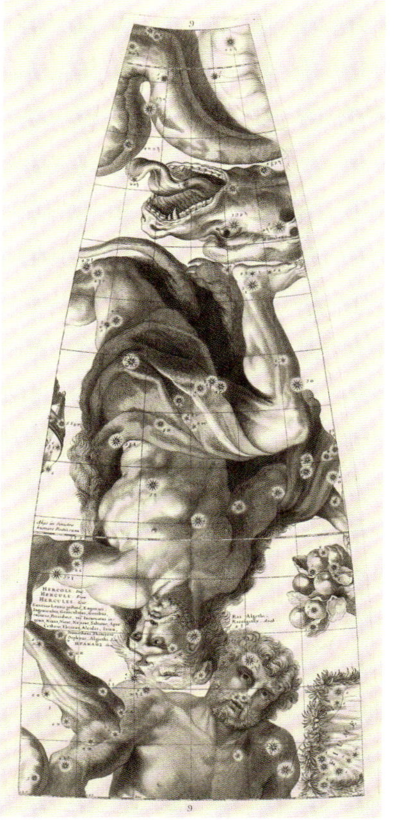

CAT. NO. 32

CAT. NO. 34

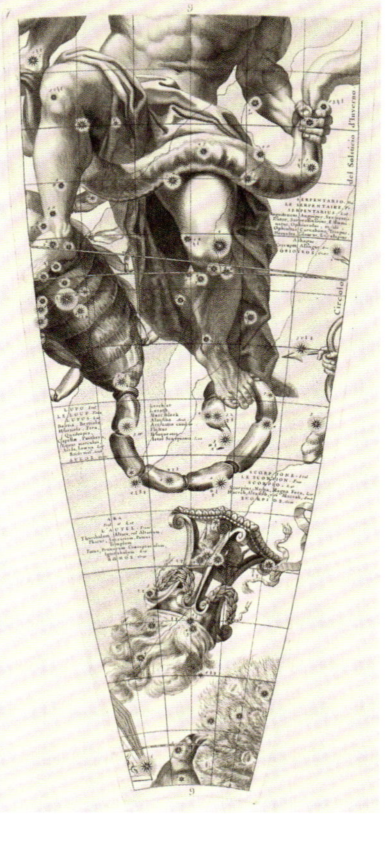

CAT. NO. 33

CAT. NO. 35

CAT. NO. 36 Vincenzo Maria Coronelli, *Set of Gores for a Nine-inch Celestial Globe*, 1700

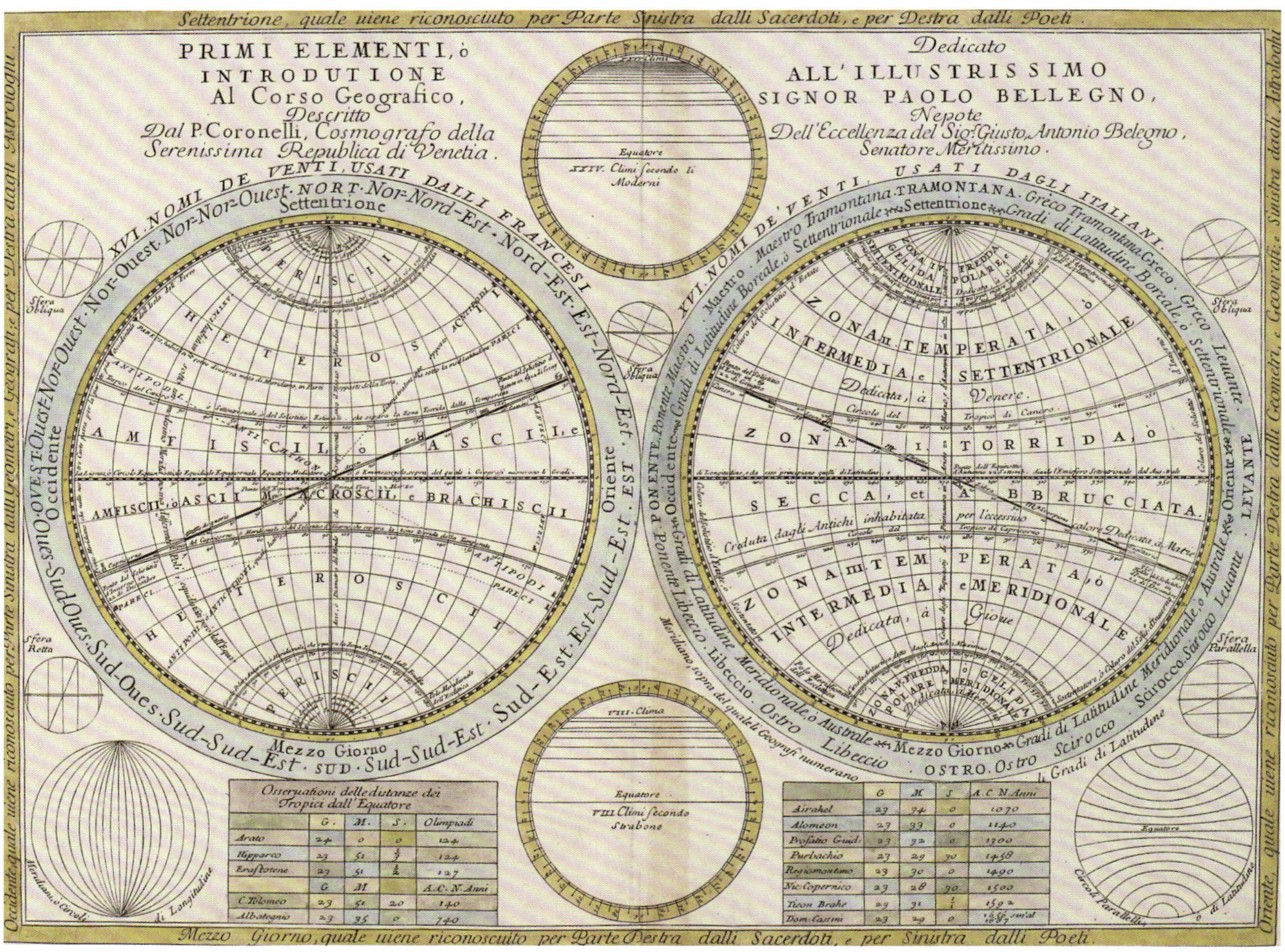

CAT. NO. 37 Vincenzo Maria Coronelli, *Primi Elementi*, 1690

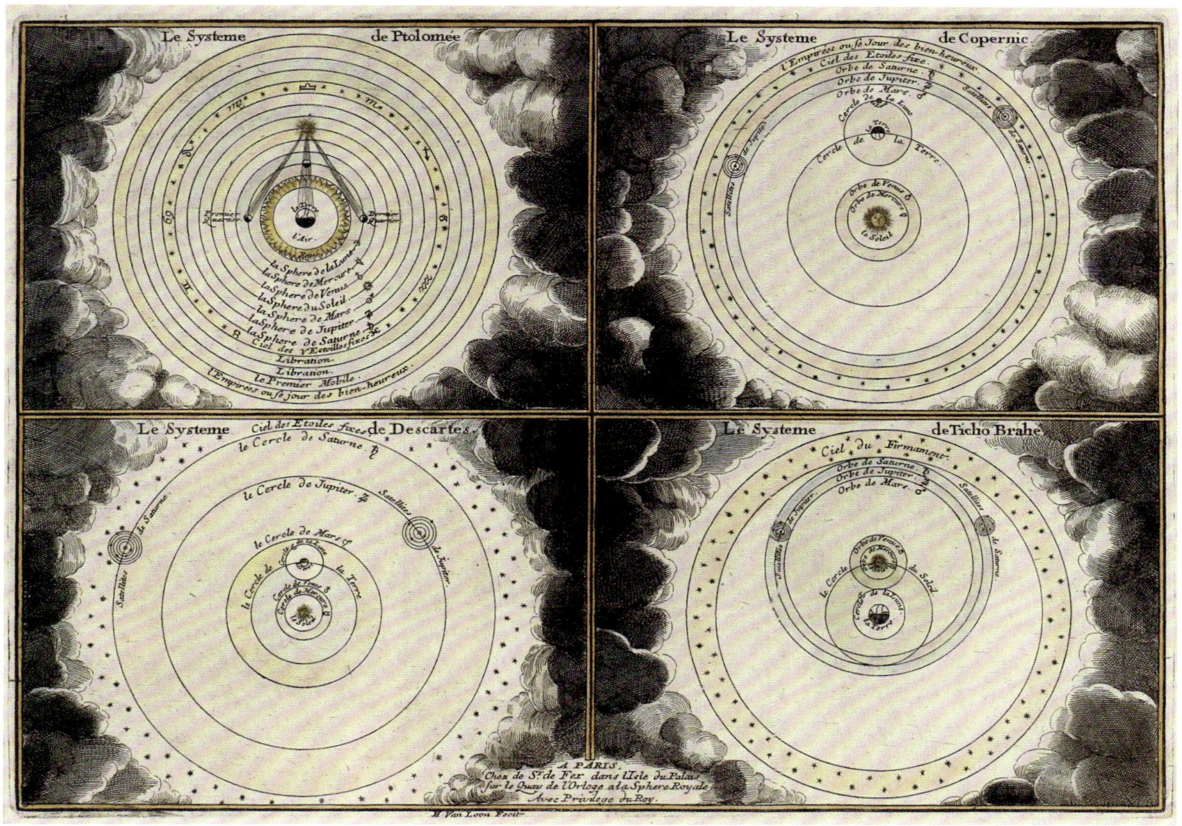

CAT. NO. 38 Nicolas de Fer with H. van Loon, engraver, *Four Systems of Cosmology* from *Atlas Historique*, 1705

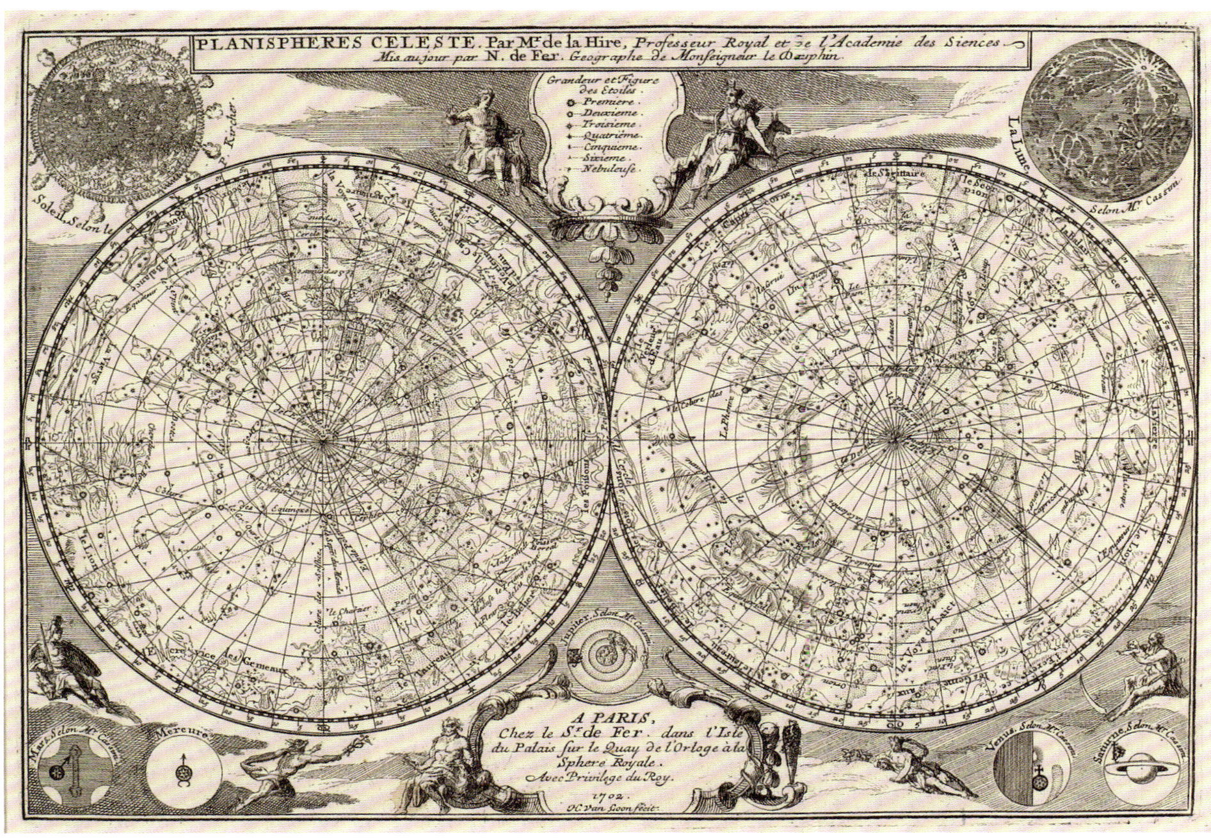

CAT. NO. 39 Nicolas de Fer with H. van Loon, engraver, *Planispheres Celeste, by de la Hire* from *L'Atlas Curieux*, 1702

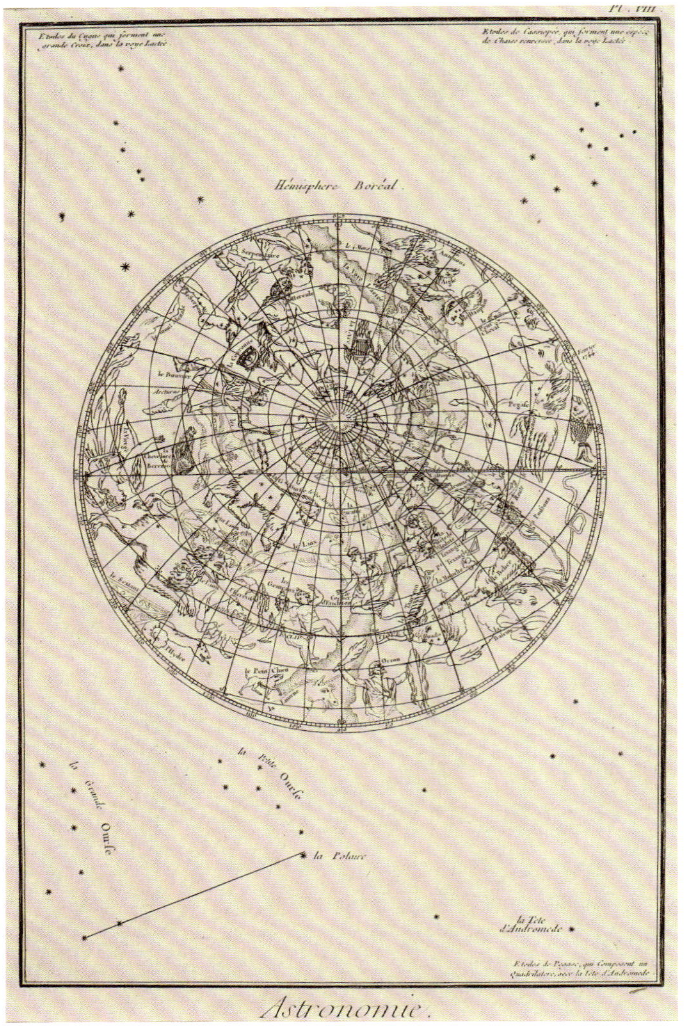

CAT. NO. 40 Denis Diderot, *Hemisphere Boreal* from *Encyclopédie ou Dictionnaire Raisonné des Sciences des Arts et des Métiers,* 1751–1780

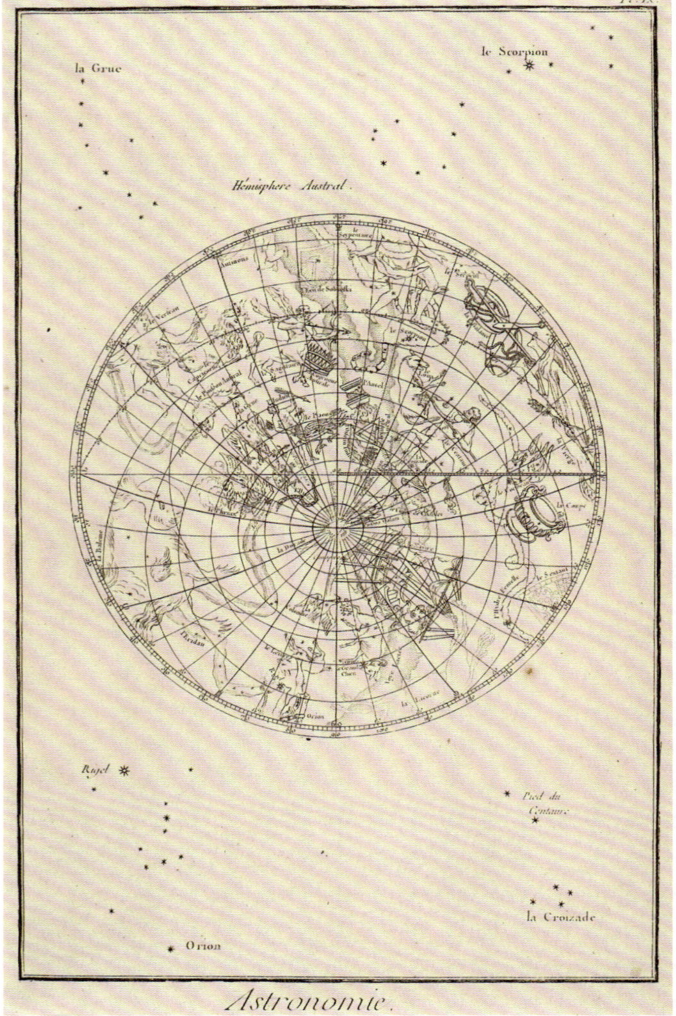

CAT. NO. 41 Denis Diderot, *Hemisphere Austral* from *Encyclopédie ou Dictionnaire Raisonné des Sciences des Arts et des Métiers,* 1751–1780

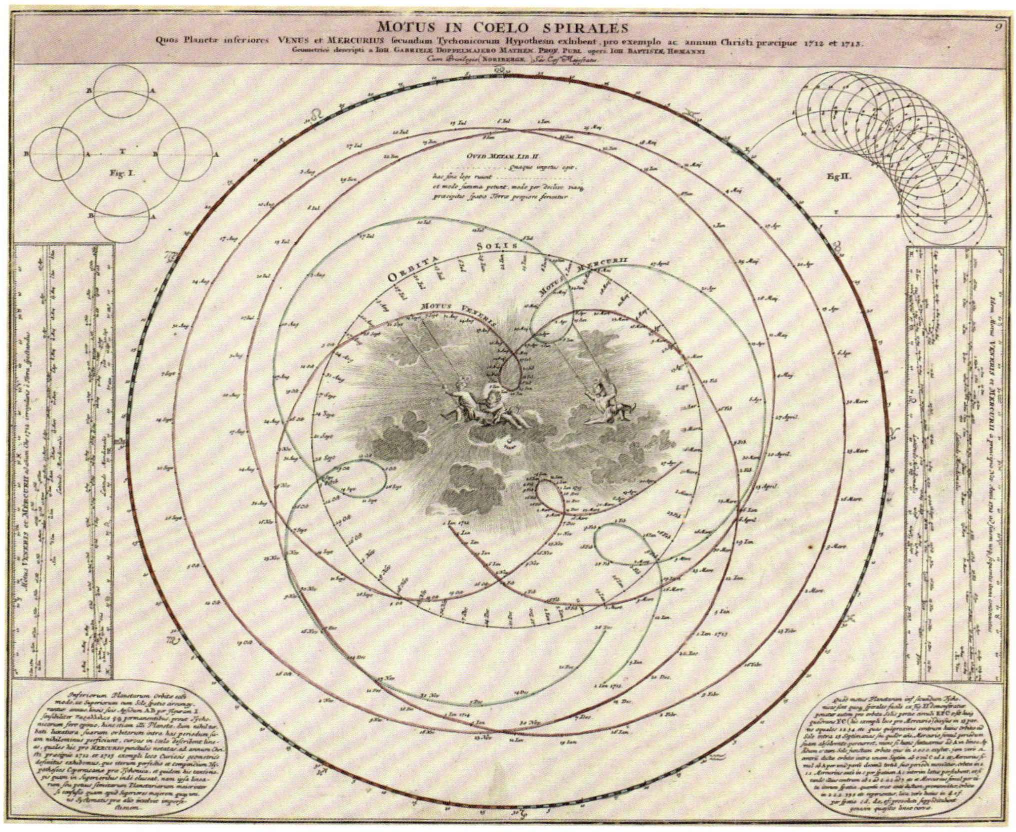

CAT. NO. 42 Johann Gabriel Doppelmayer and Johann Baptist Homann, *Motus in Coelo Spirales*, c. 1720

CAT. NO. 43 Johann Gabriel Doppelmayer and Johann Baptist Homann, *Phaenomena*, c. 1720

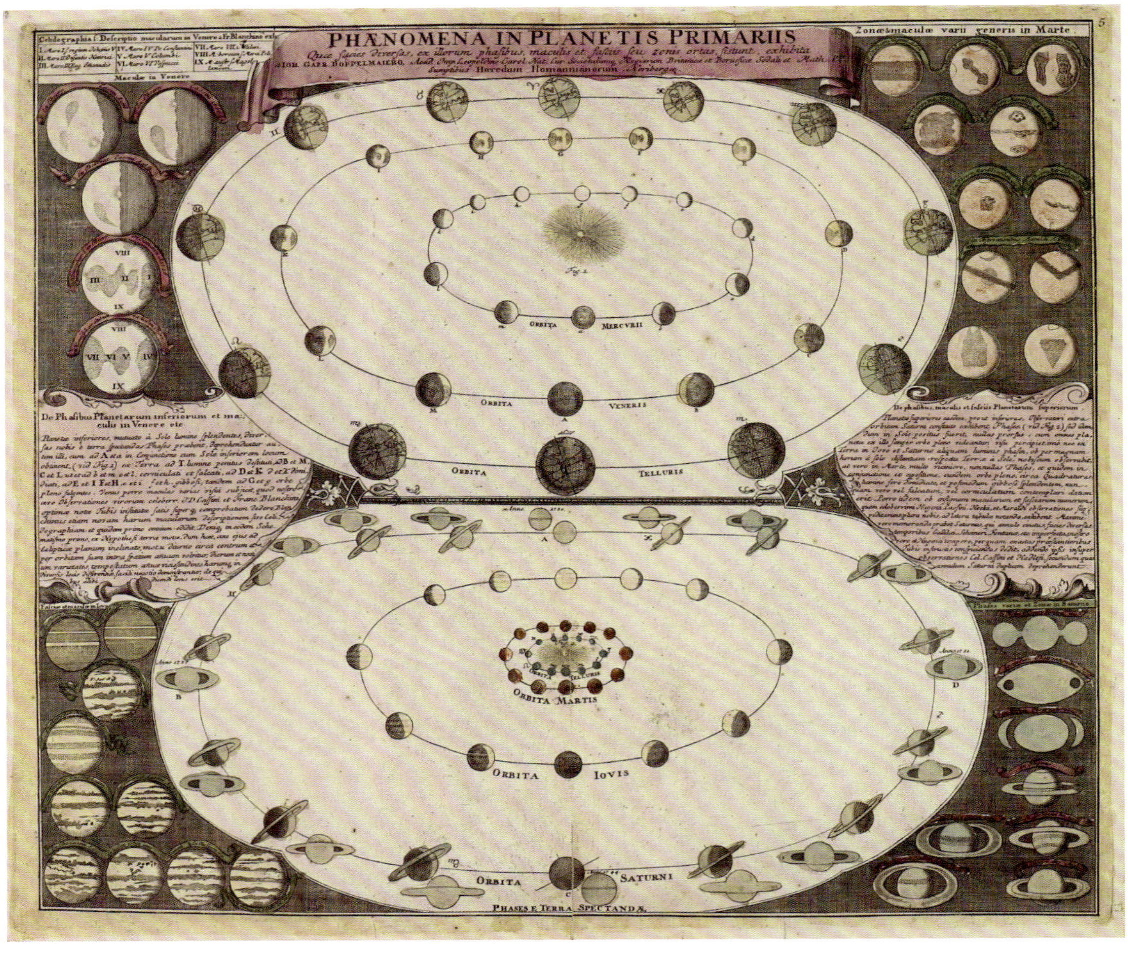

CAT. NO. 44 Johann Gabriel Doppelmayer, *Phaenomena in Planetis Primariis*, c. 1720

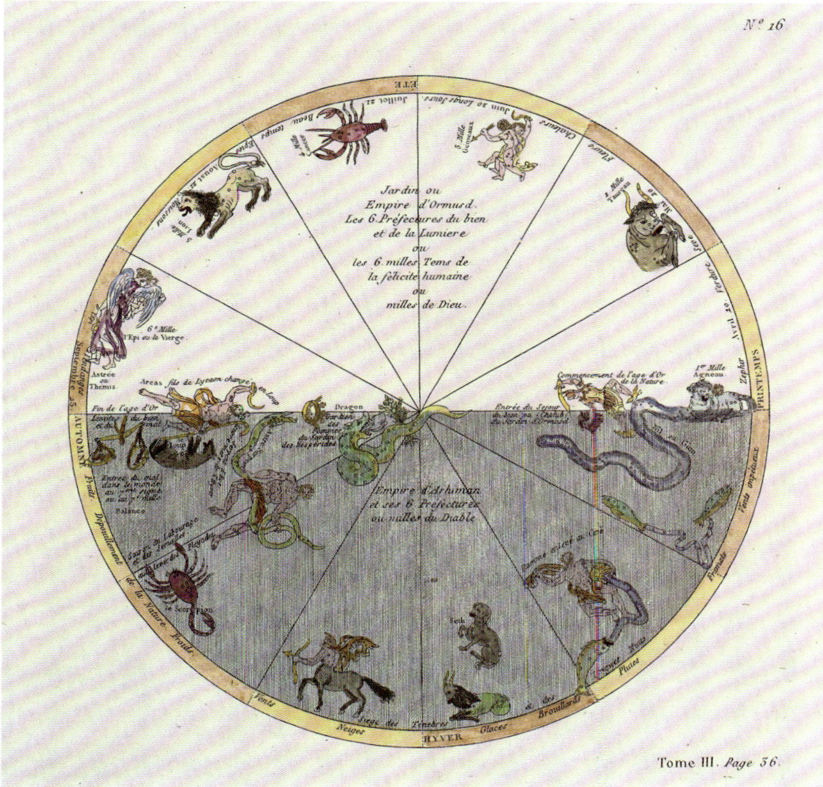

CAT. NO. 45 Charles-François Dupuis, *Jardin d'Ormusd, ou les six milles de Dieu, et les six Milles du Diable, ou l'Empire d'Ahriman* from *l'Origine de tous les cultes ou religion universelle*, 1795

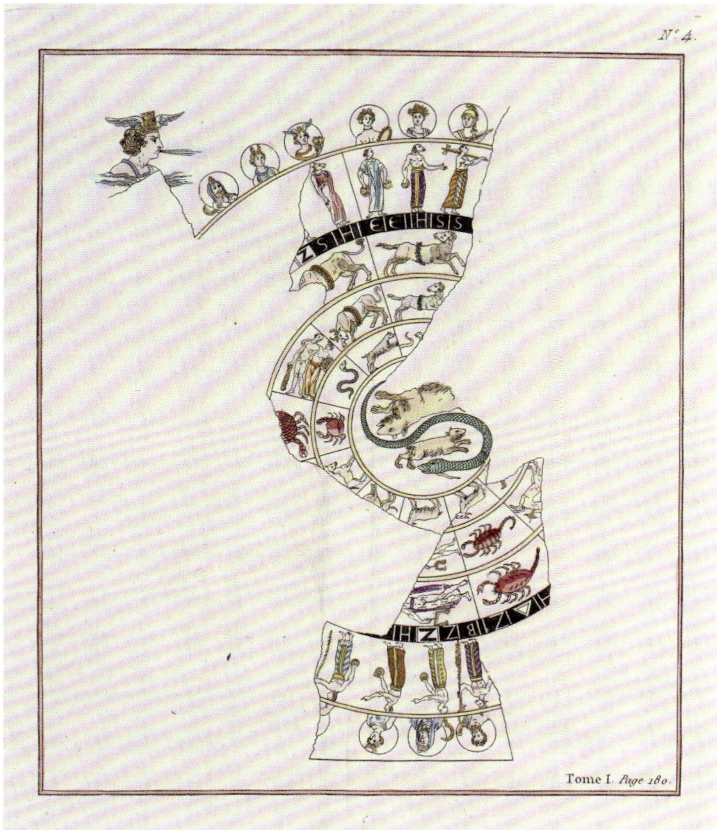

CAT. NO. 46 Charles-François Dupuis, *Planisphere astrologique de style Egyptian* from *l'Origine de tous les cultes ou religion universelle*, 1795

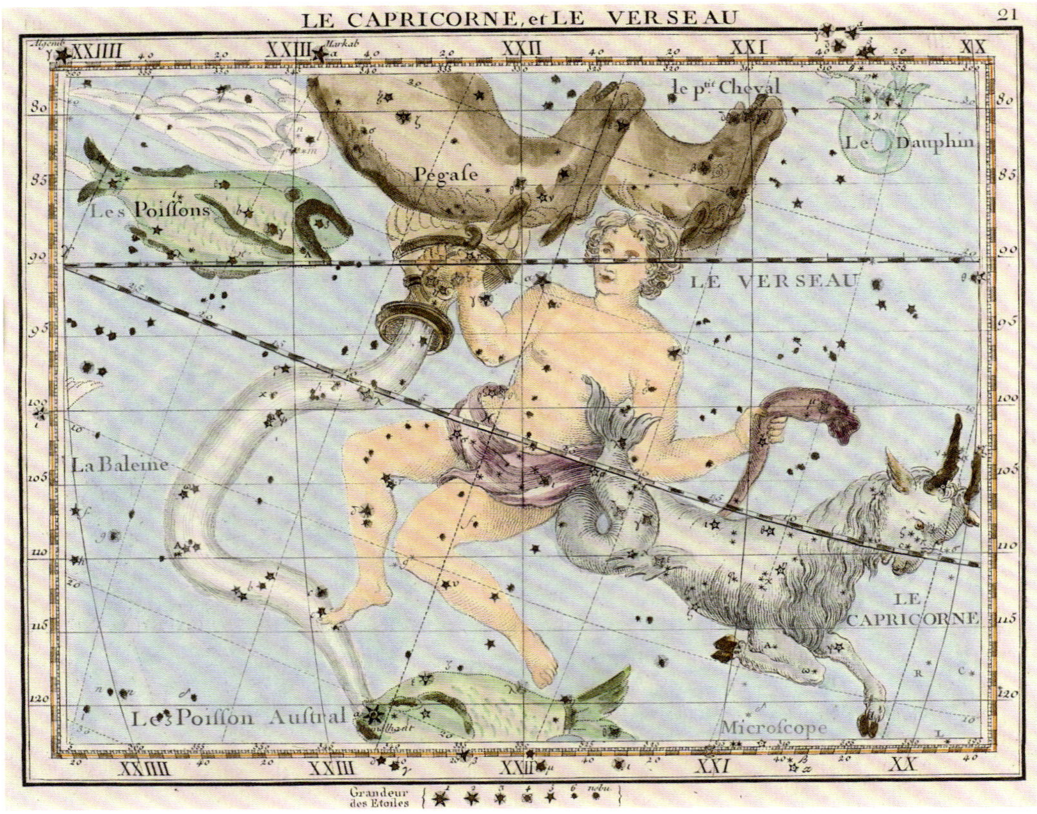

CAT. NO. 47 Jean Fortin, *Le Capricorne, et Le Verseau* from *Atlas Celeste of John Flamsteed (1646–1719)*, 1776

CAT. NO. 48 Jean Fortin, *L' Eridan, Orion, et Le Lievre* from *Atlas Celeste of John Flamsteed (1646–1719)*, 1776

CAT. NO. 49 Jean Fortin, *Le Bouvier, Les Lévriers, La Chevelute de Berenice* from *Atlas Celeste of John Flamsteed (1646–1719)*, 1776

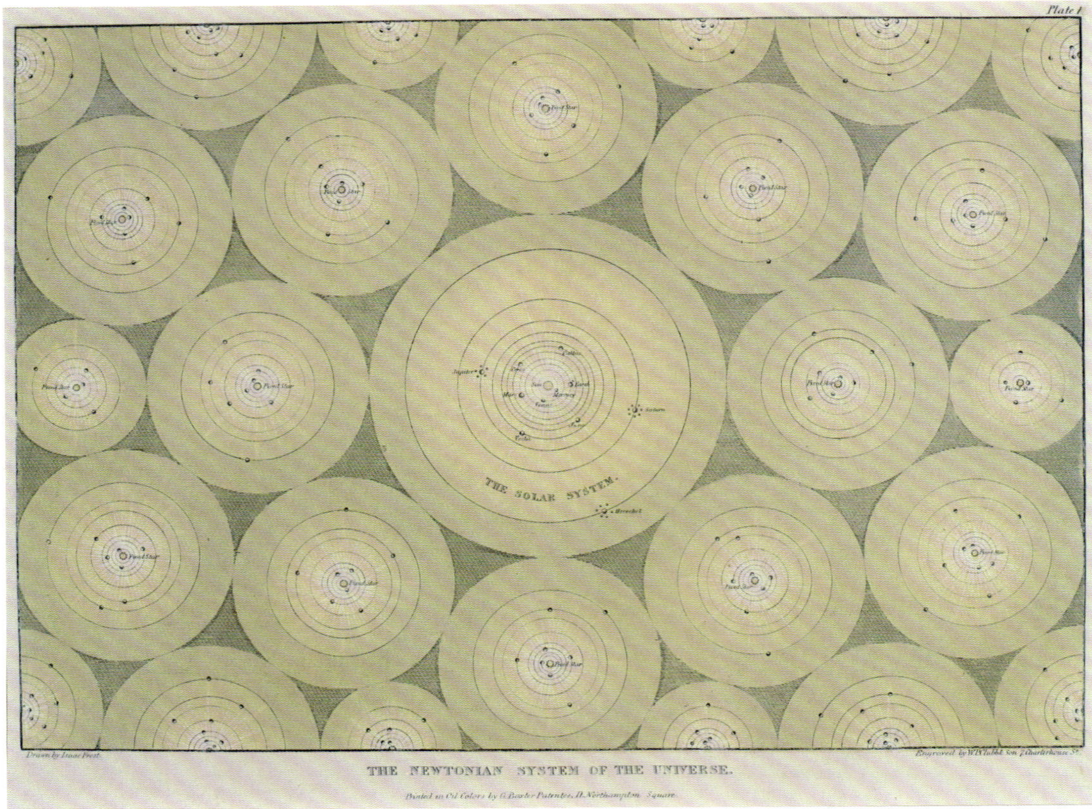

CAT. NO. 50A Isaac Frost with W. P. Clubb & Son, engravers, *The Newtonian System of the Universe*, from *Two Systems of Astronomy*, a Muggeltonian publication with oil printing by G. Baxter, 1846

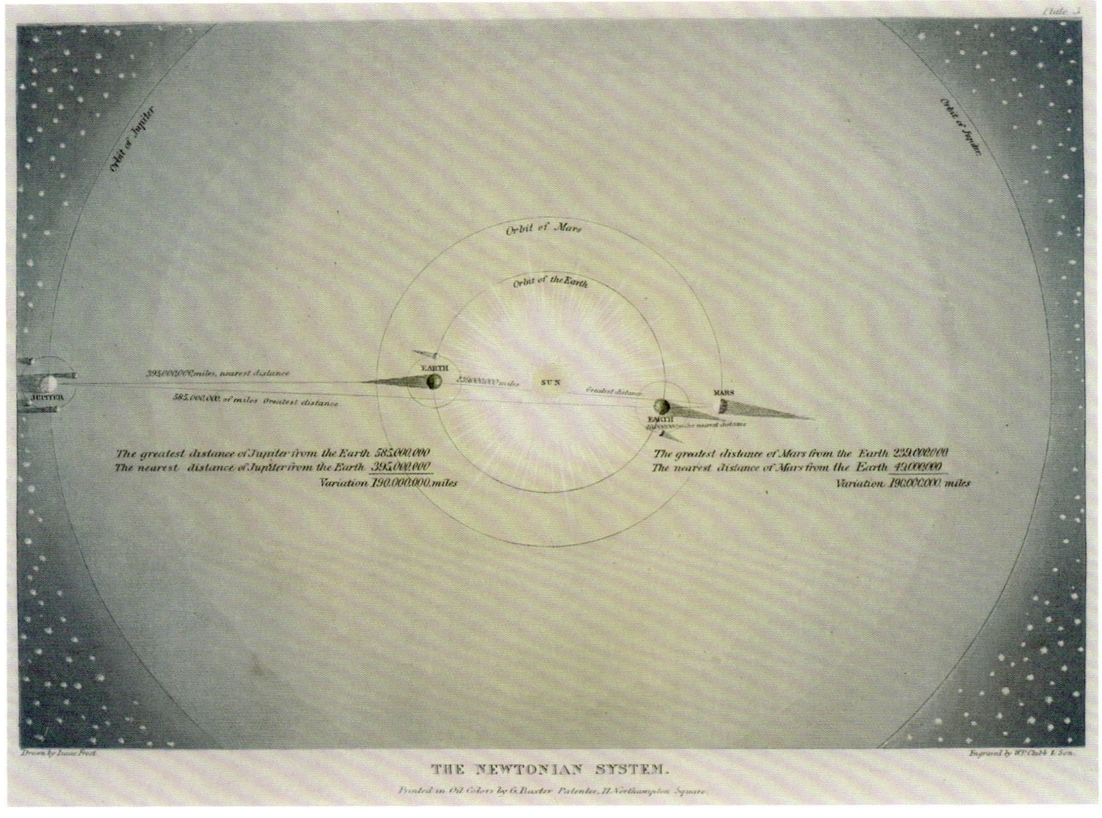

CAT. NO. 50B Isaac Frost with W. P. Clubb & Son, engravers, *The Newtonian System*

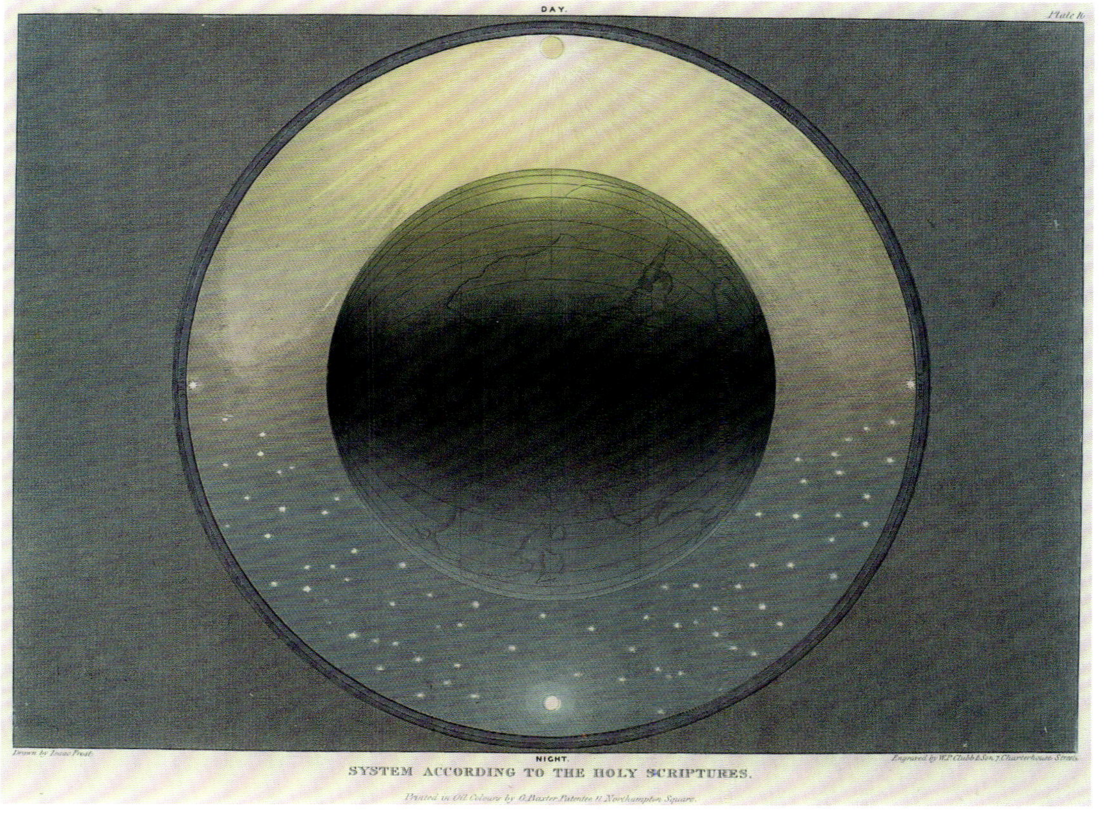

CAT. NO. 50C Isaac Frost with W. P. Clubb & Son, engravers, *System According to the Holy Scriptures*

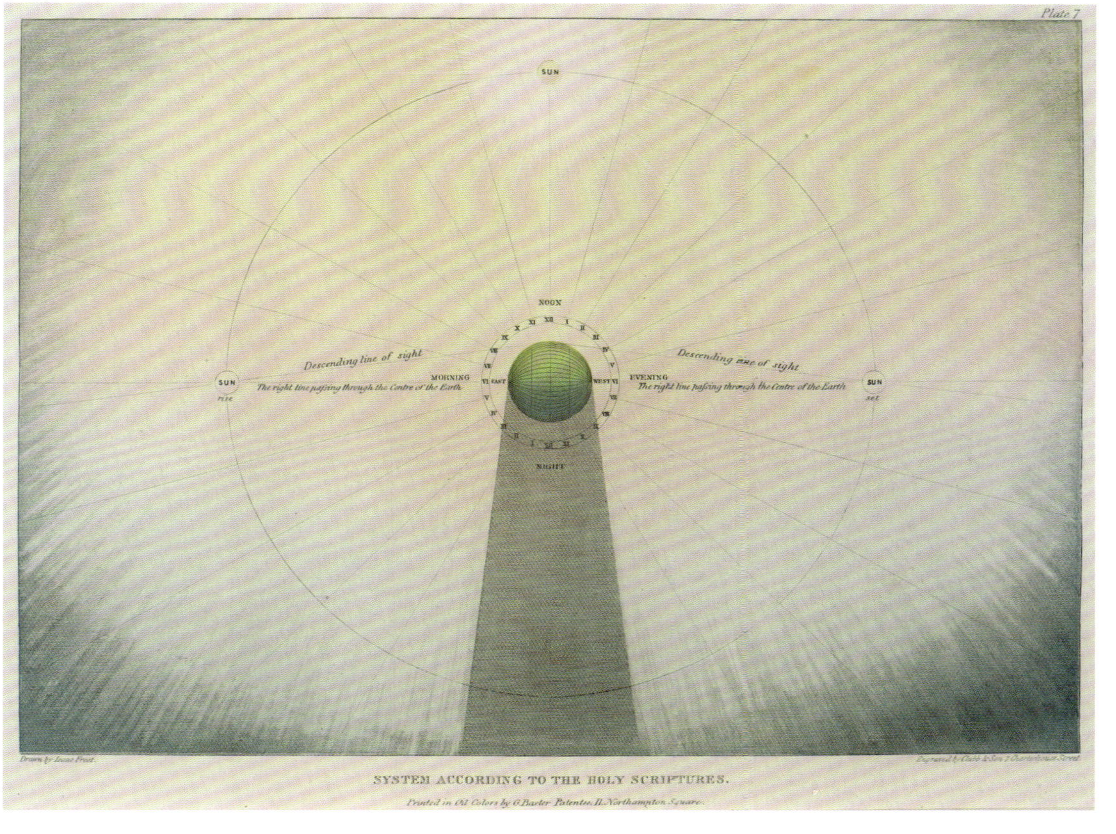

CAT. NO. 50D Isaac Frost with W. P. Clubb & Son, engravers, *System According to the Holy Scriptures*

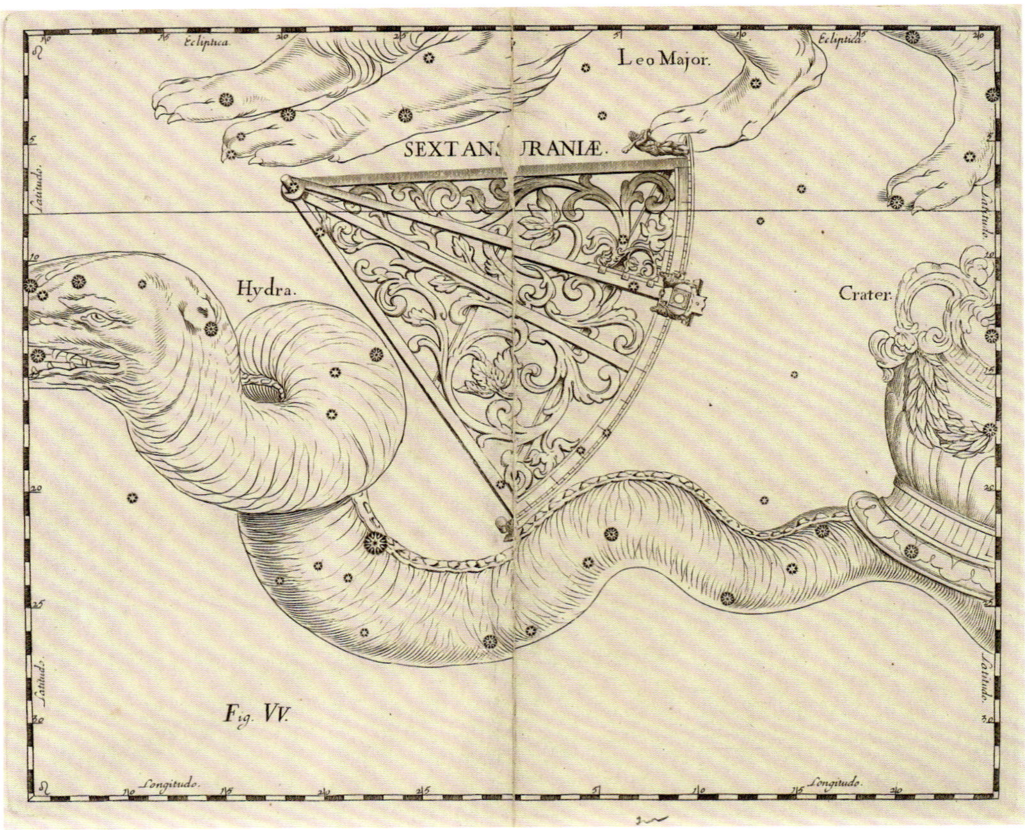

CAT. NO. 51 Johannes Hevelius, *Sextant* from *Firmamentum Sobiescianum*, 1687–1690

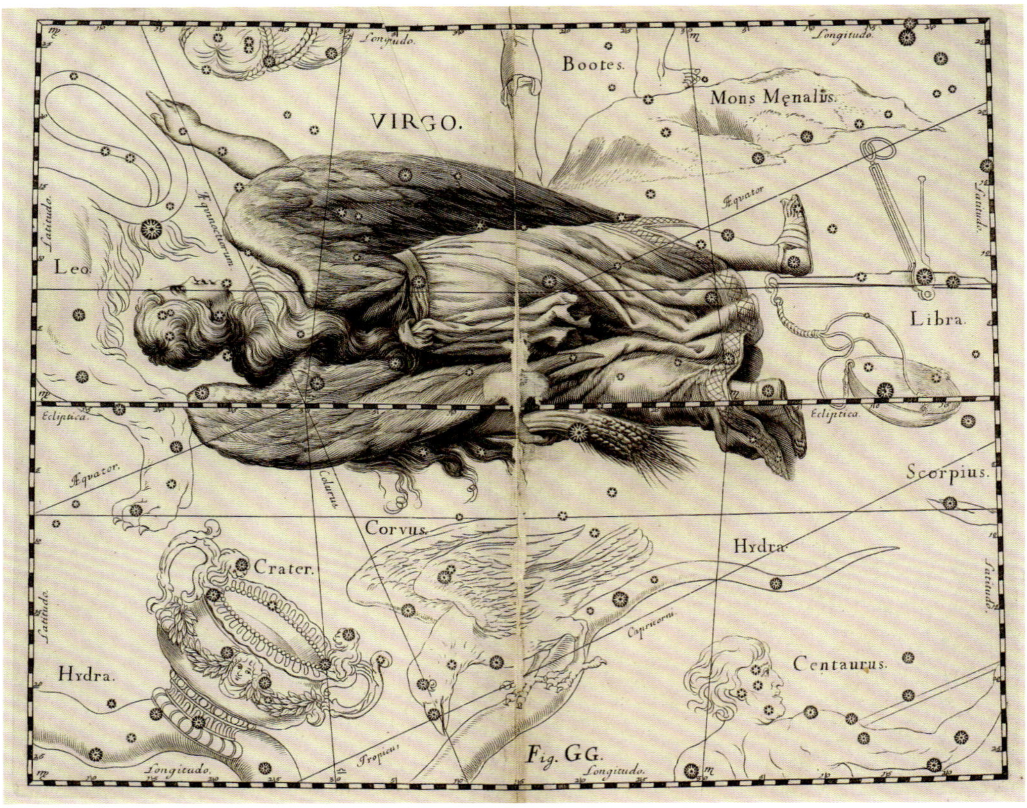

CAT. NO. 52 Johannes Hevelius, *Virgo* from *Firmamentum Sobiescianum*, 1687–1690

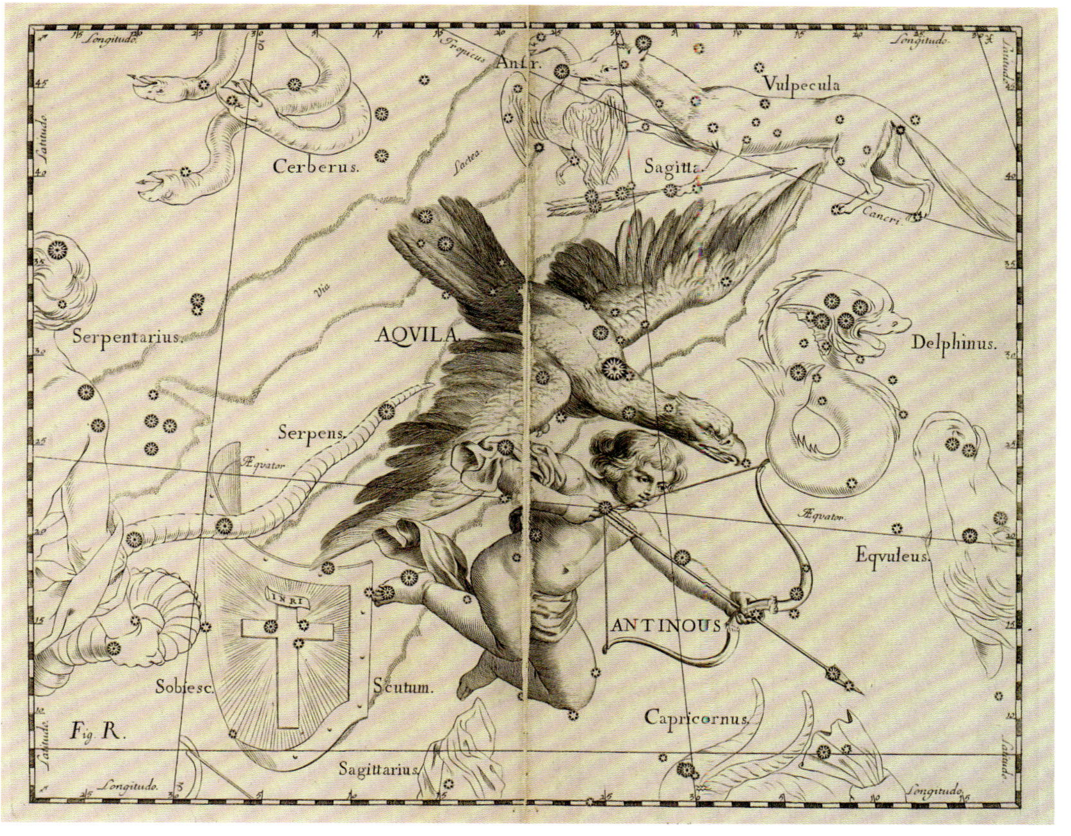

CAT. NO. 53 Johannes Hevelius, *Aquila and Antinous* from *Firmamentum Sobiescianum*, 1687–1690

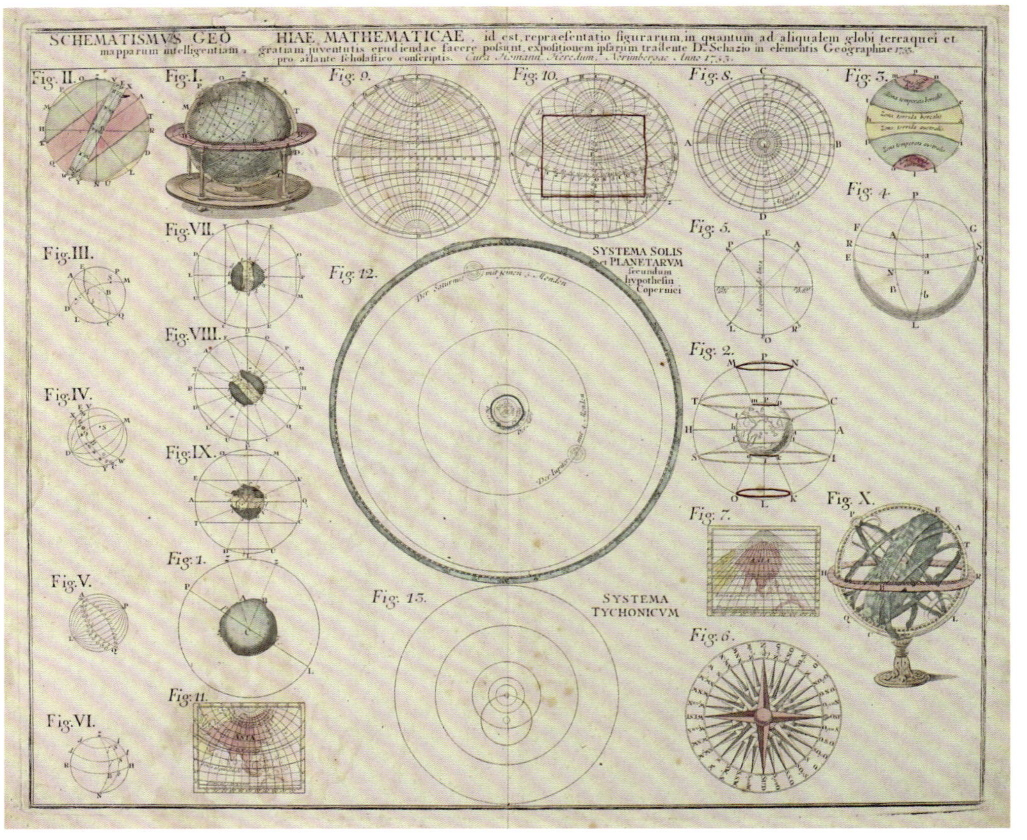

CAT. NO. 54 Johann Baptist Homann, *Schematismus Geographiae Mathematicae*, 1753

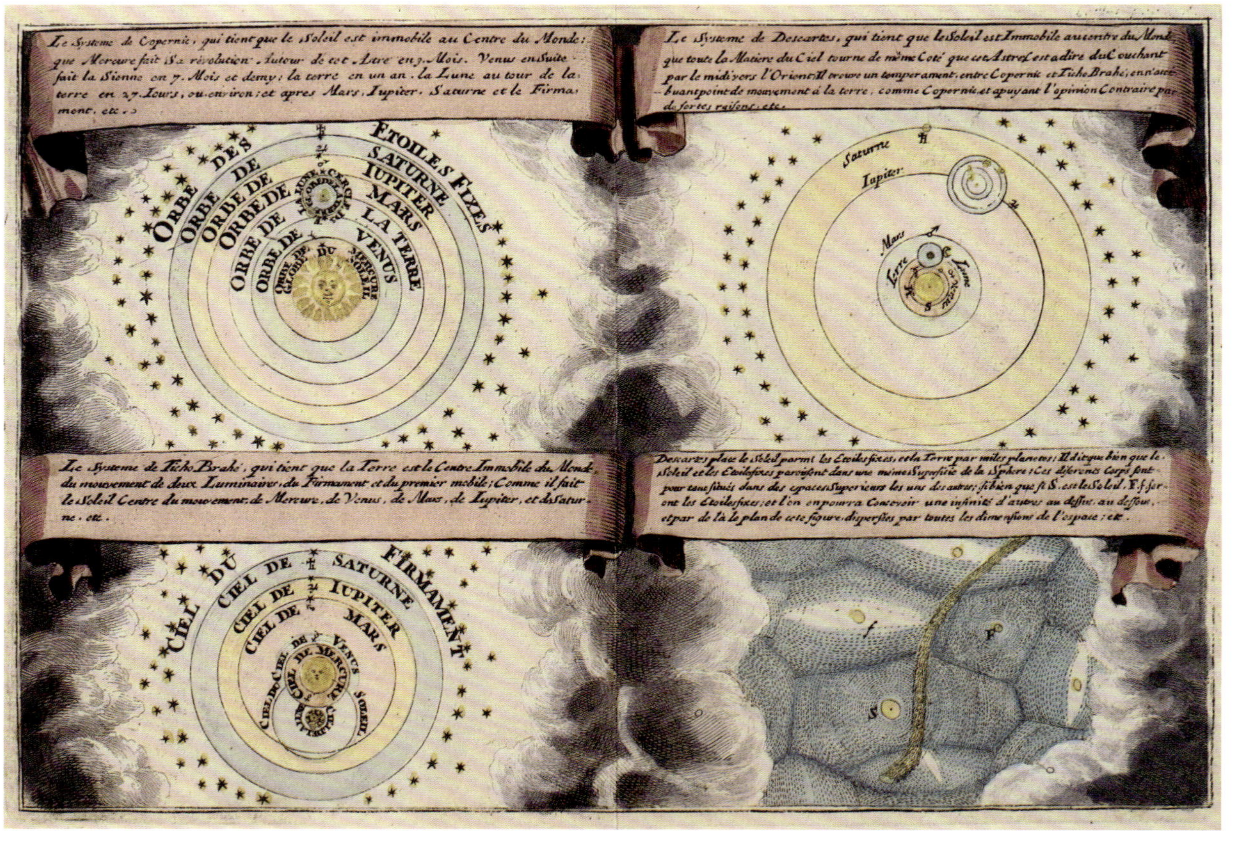

CAT. NO. 55 Alexis Hubert Jaillot, *Four Systems of Cosmology*, c. 1690

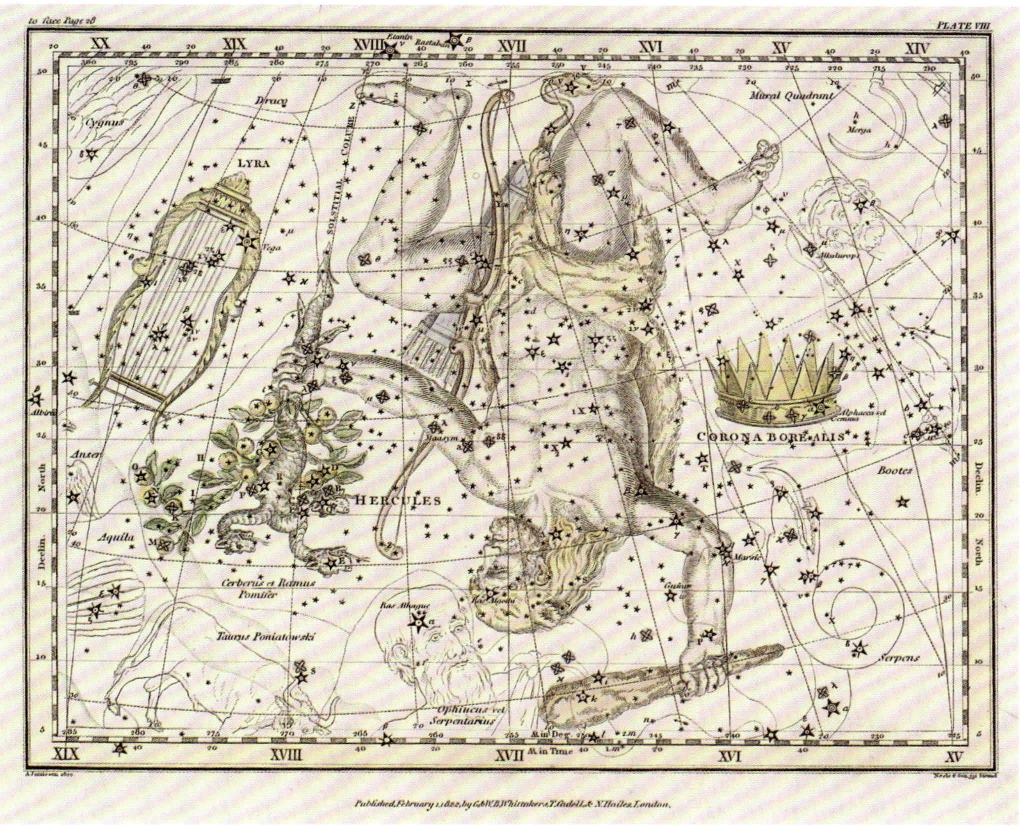

CAT. NO. 56 Alexander Jamieson with Neele & Son, engravers, *Lyra, Hercules, Corona Borealis* from *A Celestial Atlas*, published by G. & W. B. Whittakers, T. Cadell & N. Hailes, 1822

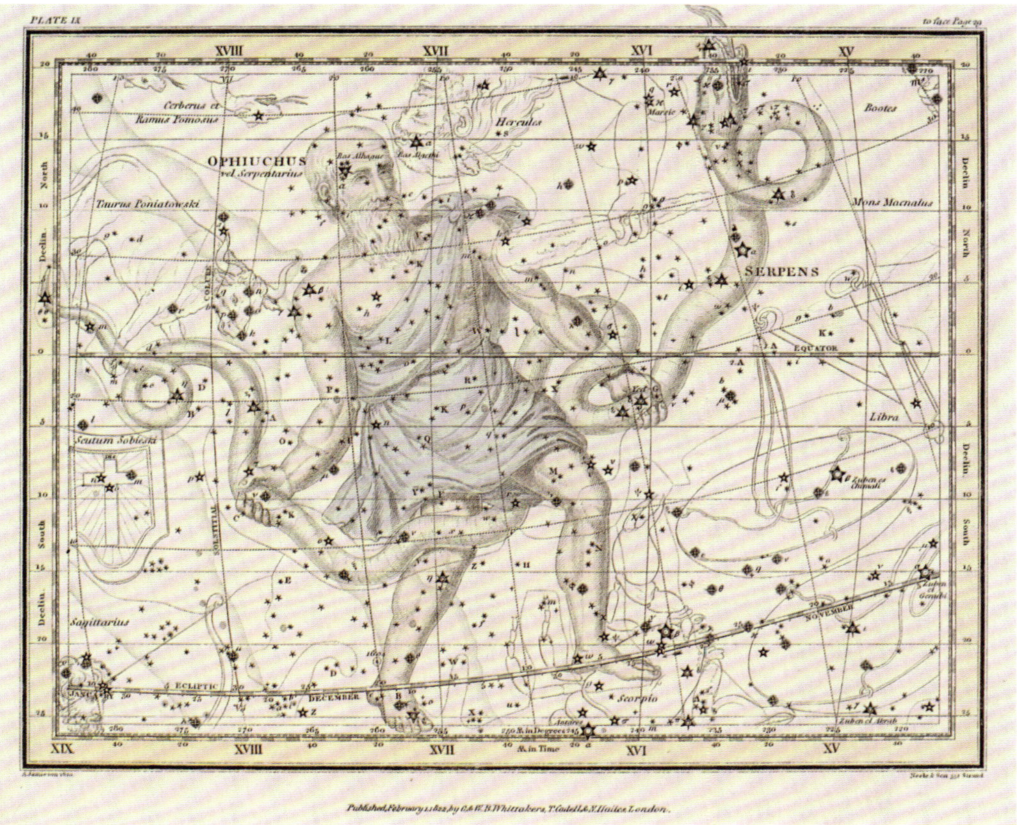

CAT. NO. 57 Alexander Jamieson with Neele & Son, engravers, *Ophiuchus, Serpens* from *A Celestial Atlas*, published by G. & W. B. Whittakers, T. Cadell & N. Hailes, 1822

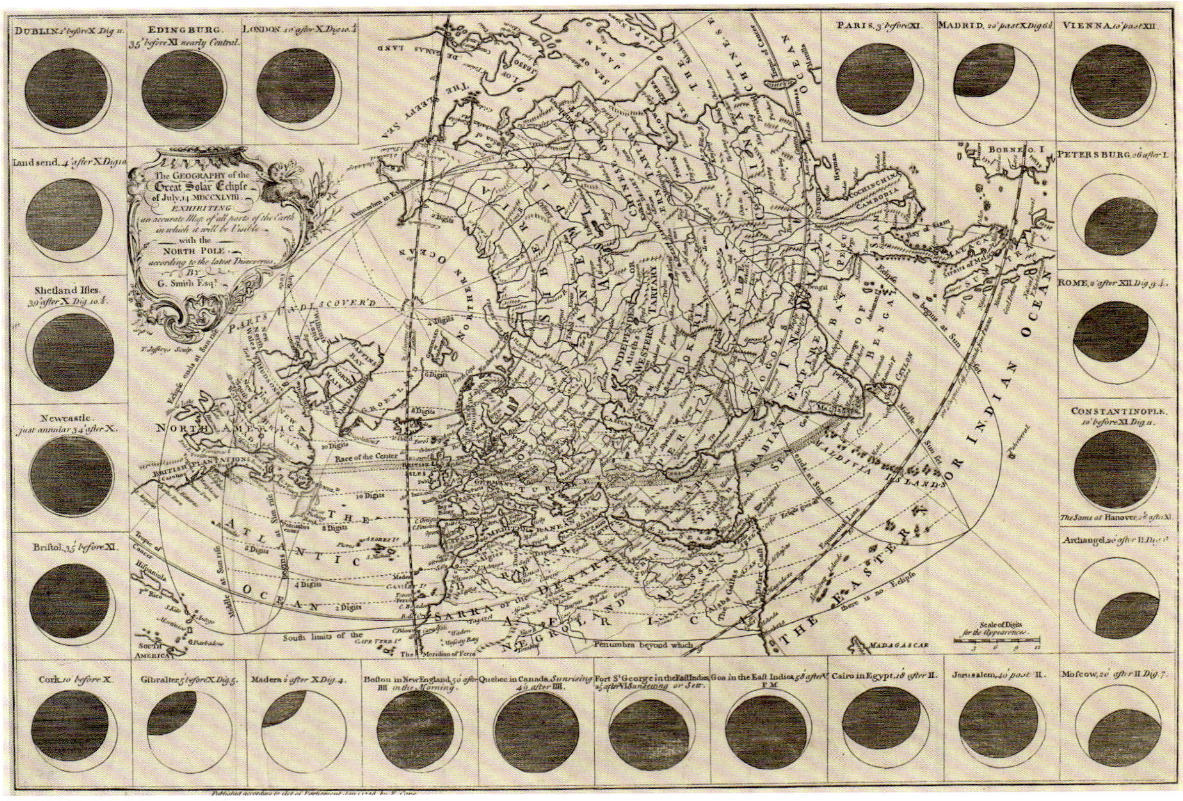

CAT. NO. 58 T. Jeffreys, *The Geography of the Great Solar Eclipse of July 14, 1748*, published by E. Cave, 1748

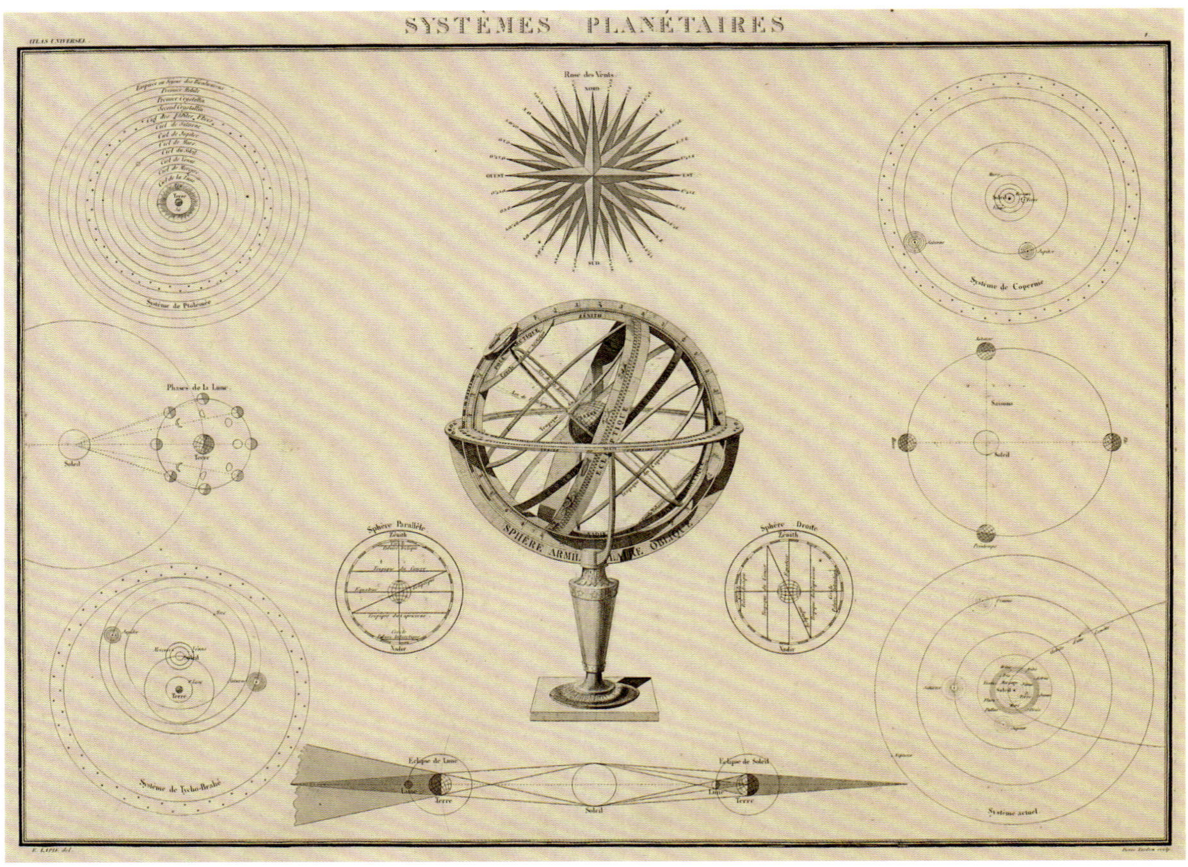

CAT. NO. 59 E. Lapie with Pierre-Alexandre Tardieu, engraver, *Systèmes Planétaires* from *Atlas Universel*, 1838

CAT. NO. 60 Allain Manesson Mallet, *Eclipse of the Moon* from *Description de l'Univers*, 1683

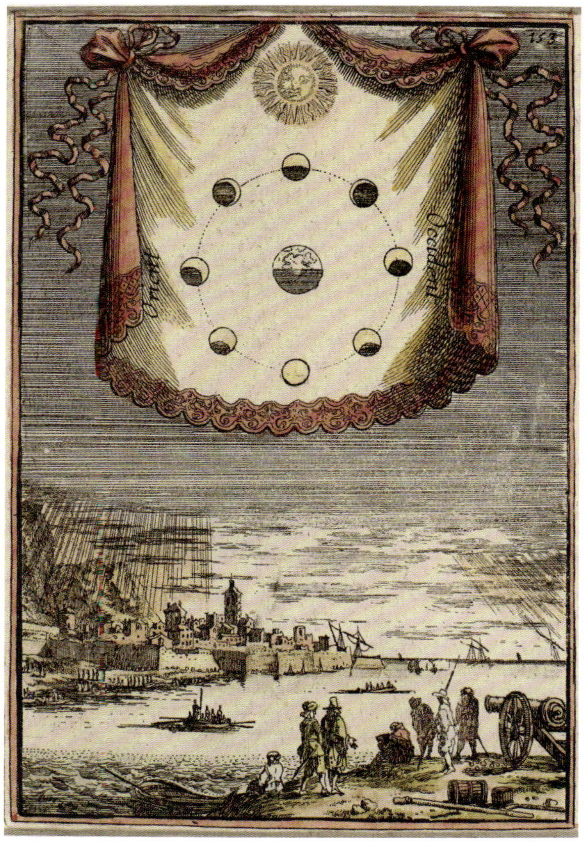

CAT. NO. 61 Allain Manesson Mallet, *Phases of the Moon* from *Description de l'Univers*, 1683

CAT. NO. 62 Allain Manesson Mallet, *Ursa Major, Ursa Minor & Cetus* from *Description de l'Univers*, 1683

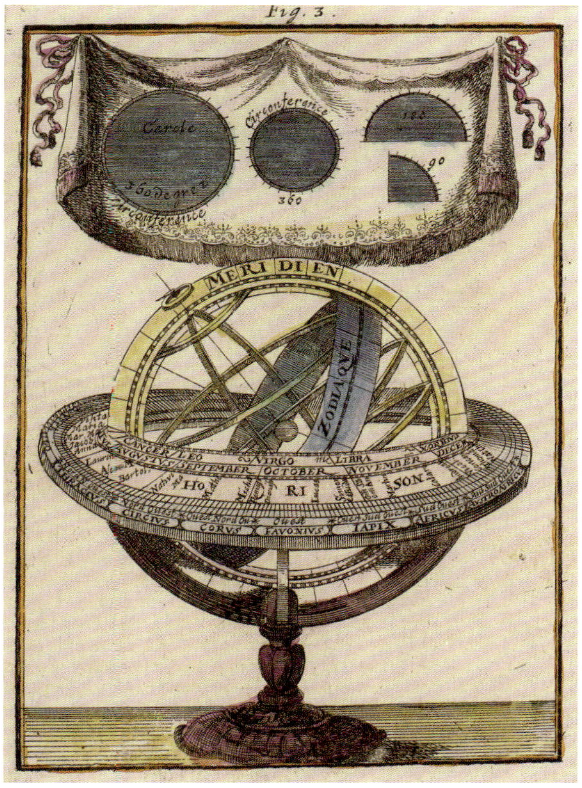

CAT. NO. 63 Allain Manesson Mallet, *Armillary Sphere* from *Description de l'Univers*, 1683

CAT. NO. 64 Allain Manesson Mallet, *Types of Eclipses* from *Description de l'Univers*, 1683

CAT. NO. 65 Allain Manesson Mallet, *Copernican System* from *Description de l'Univers*, 1683

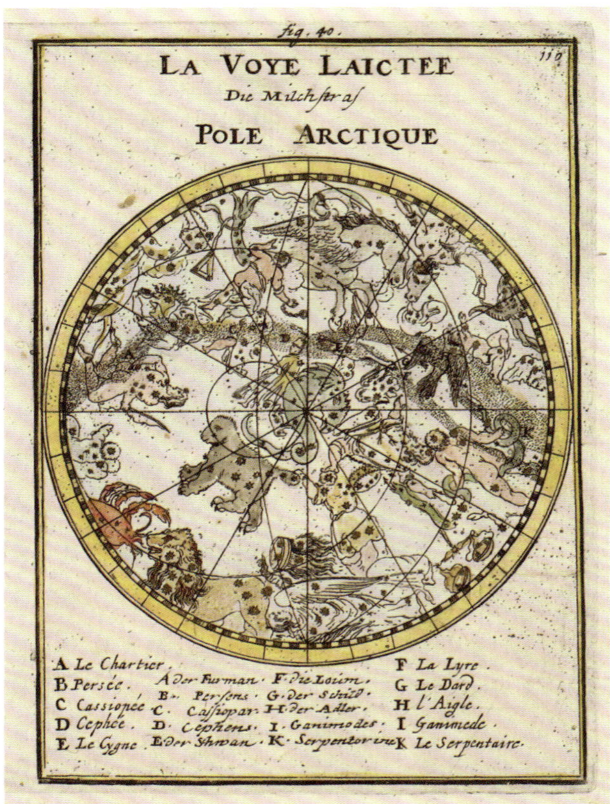

CAT. NO. 66 Allain Manesson Mallet, *Pole Arctique* from *La Voye Laictée*, 1684

CAT. NO. 67 Allain Manesson Mallet, *Pole Antarctique* from *La Voye Laictée*, 1684

CAT. NO. 68A *Sun*

CAT. NO. 68B *Moon*

CAT. NO. 68C *Mercury*

CAT. NO. 68D *Venus*

CAT. NO. 68A–G Allain Manesson Mallet, *Members of the Solar System: Sun, Moon, Mercury, Venus, Mars, Jupiter and Saturn* from *Description de l'Univers*, seventeenth century (German edition)

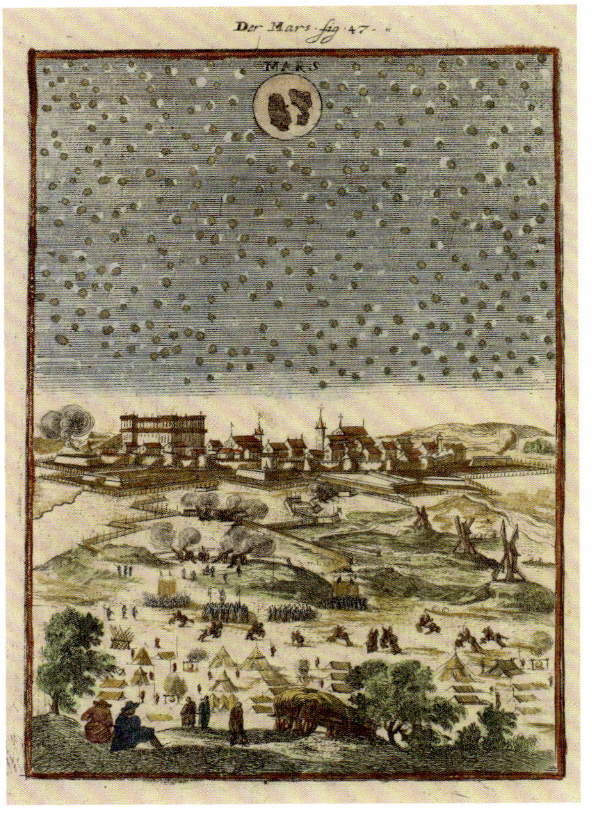

CAT. NO. 68E *Mars*

CAT. NO. 68F *Jupiter*

CAT. NO. 68G *Saturn*

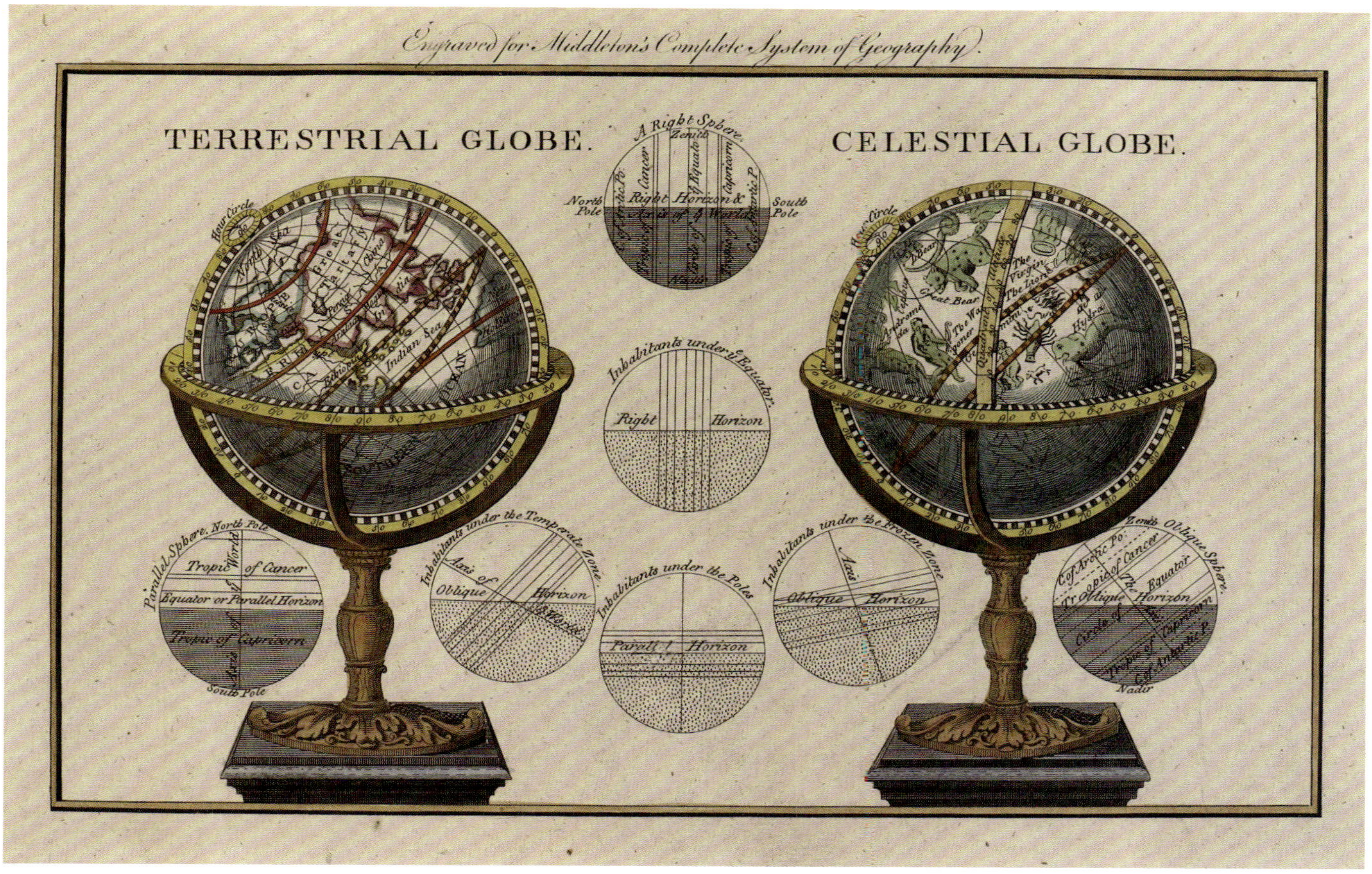

CAT. NO. 69 Charles Theodore Middleton, *Terrestrial Globe and Celestial Globe* from *Middleton's Complete System of Geography*, 1750

CAT. NO. 70 Julius Schiller, *Mary Magdalene* from *Coelum Stellatum Christianum*, 1627

CAT. NO. 71 Julius Schiller, *St. Helen* from *Coelum Stellatum Christianum*, 1627

CAT. NO. 72 Julius Schiller, *St. Joseph* from *Coelum Stellatum Christianum*, 1627

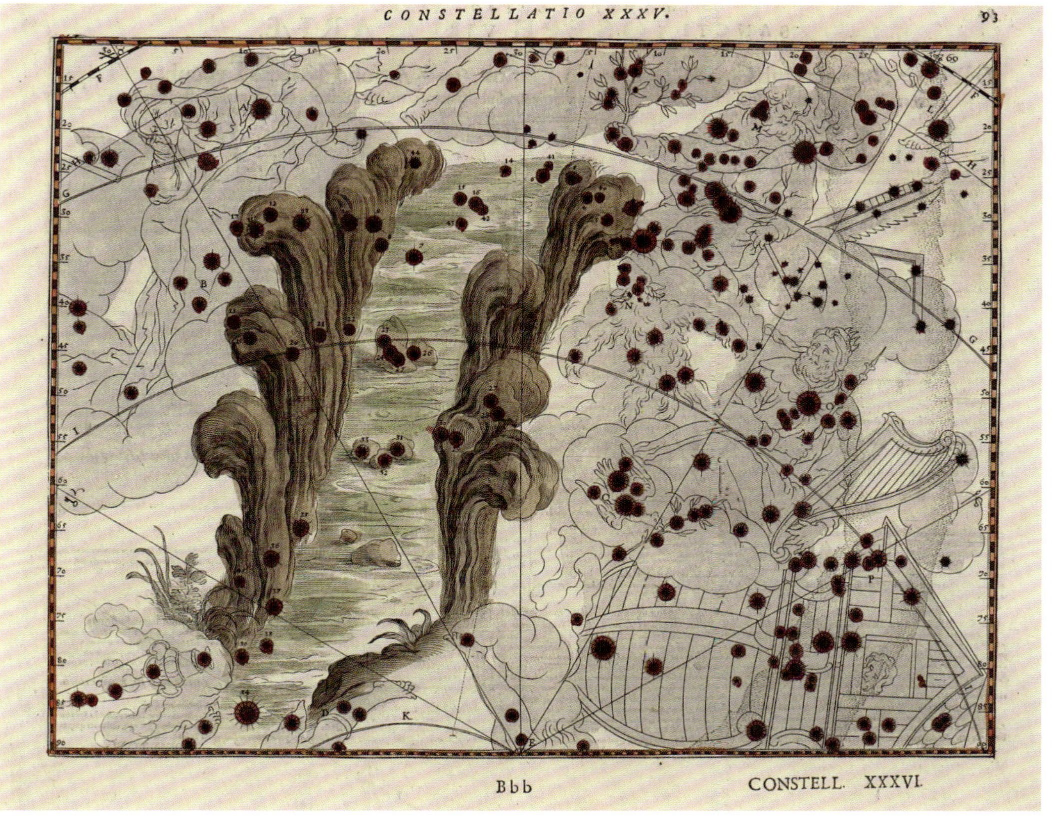

CAT. NO. 73 Julius Schiller, *Transitus Israel* from *Coelum Stellatum Christianum*, 1627

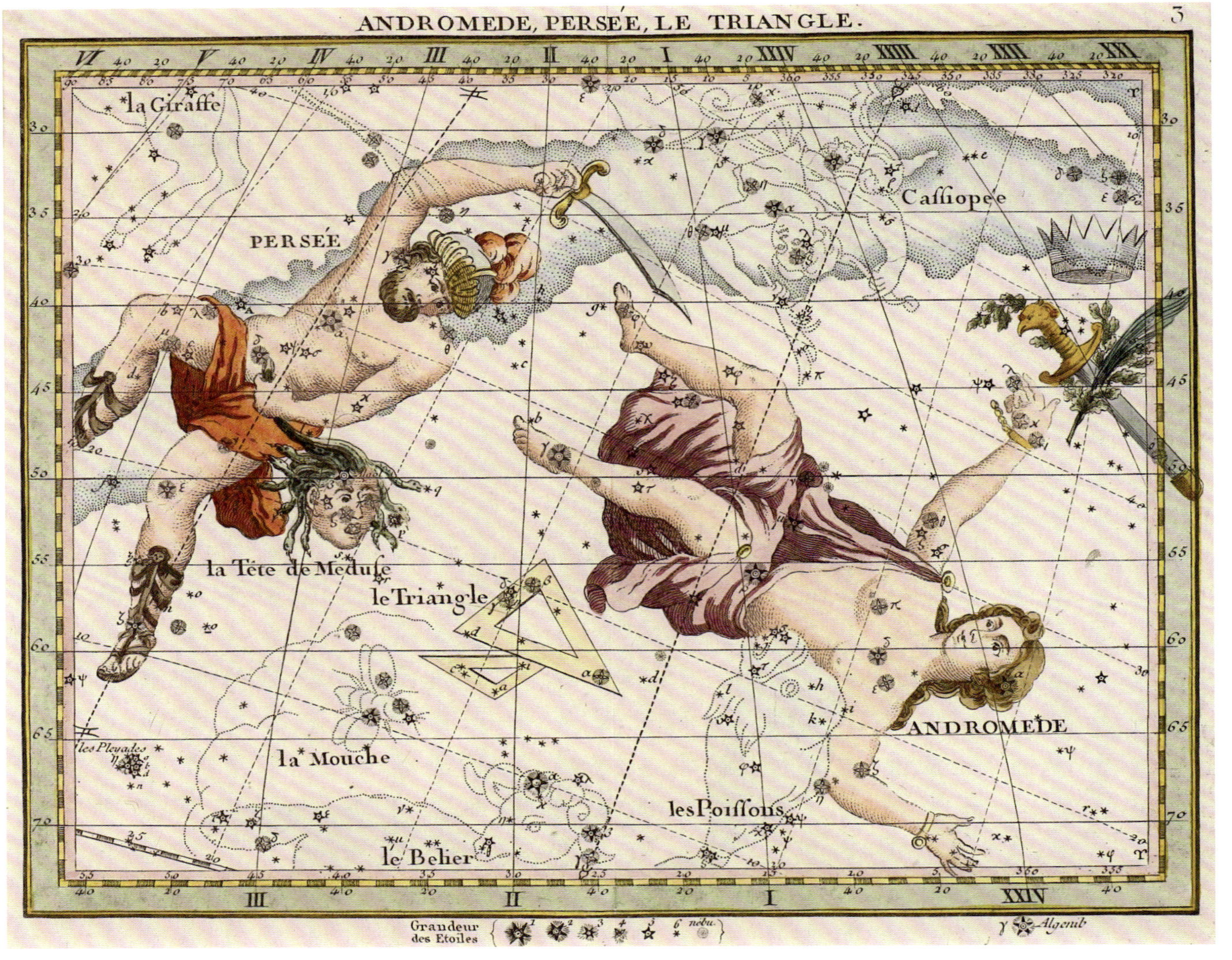

CAT. NO. 74 Unknown engraver, after John Flamsteed, *Andromède, Persée, le Triangle*, c. 1830

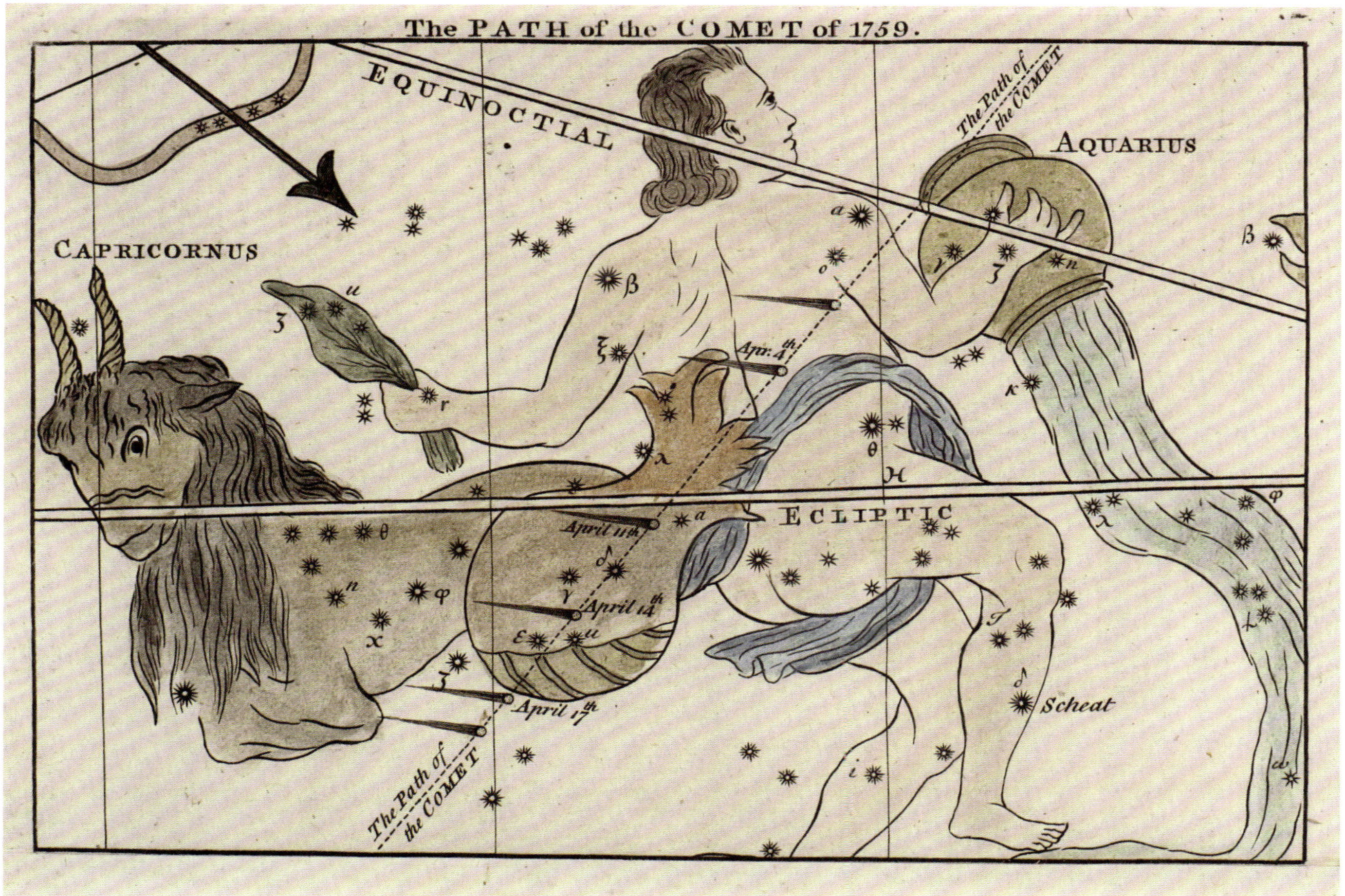

CAT. NO. 75 Unknown engraver, *The Path of the Comet of 1759*, published in the *Gentleman's Magazine*, 1759

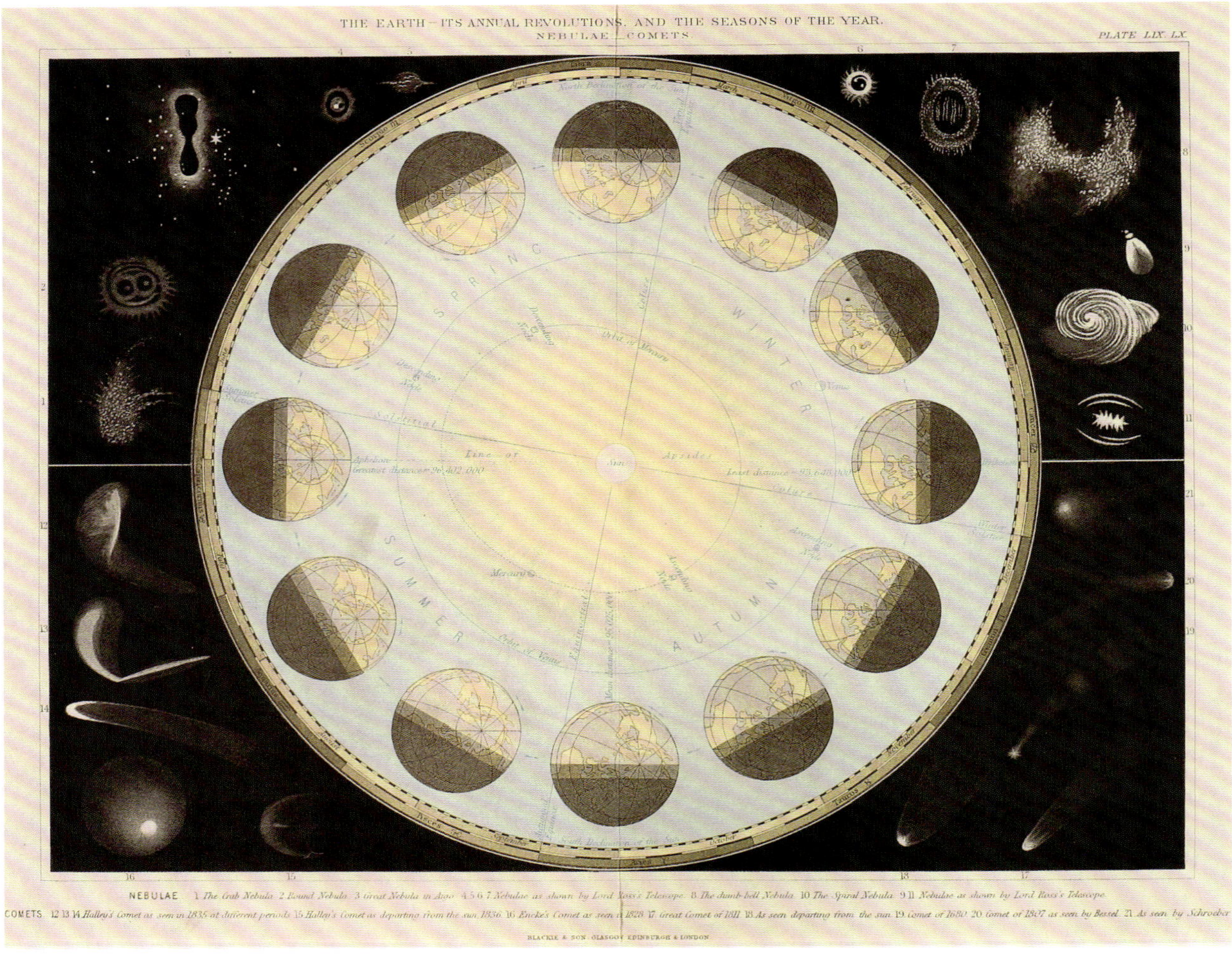

CAT. NO. 76 Unknown engraver-cartographer, *The Earth, Nebulae and Comets*, published by Blackie & Son, c. 1880

CAT. NO. 77 Unknown artist, *Constellations of the Zodiac* from an unidentified Islamic book, 1776

CAT. NO. 78 Unknown artist, *Planispheres* from an unidentified Islamic book, 1776

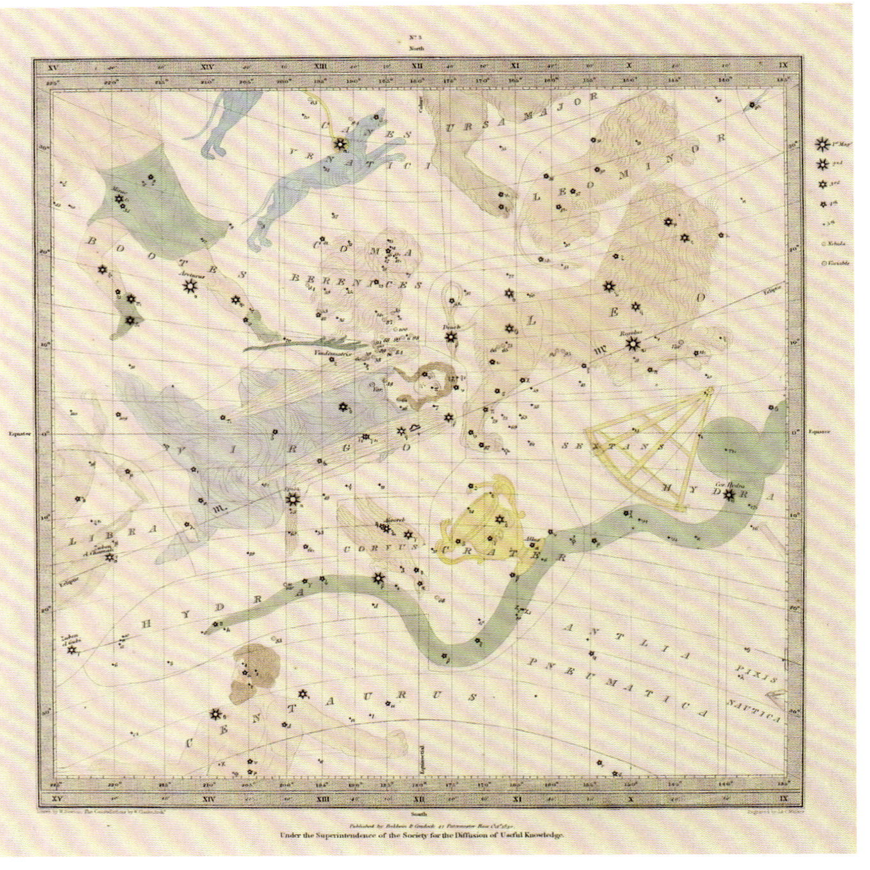

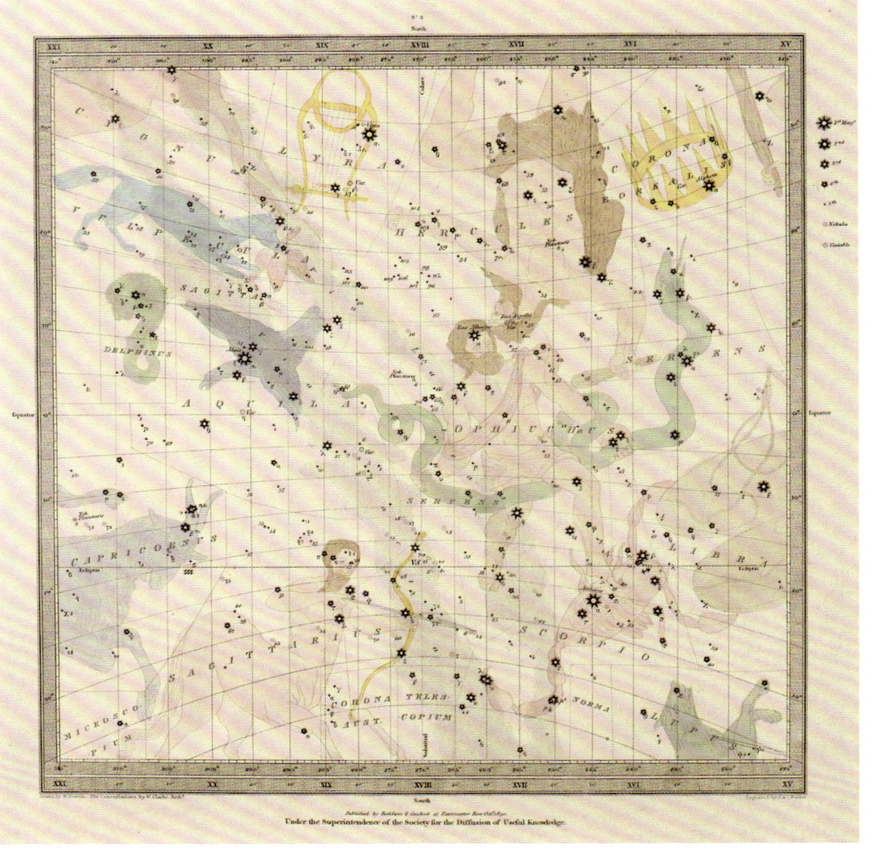

CAT. NO. 79 J. & C. Walker, drawn by W. Newton; constellations by W. Clarke, *The Constellations: No. 3, Equinoctial* and *The Constellations: No. 4, Solstitial* published by Baldwin & Cradock under the Superintendence of the Society for the Diffusion of Useful Knowledge, London, 1830

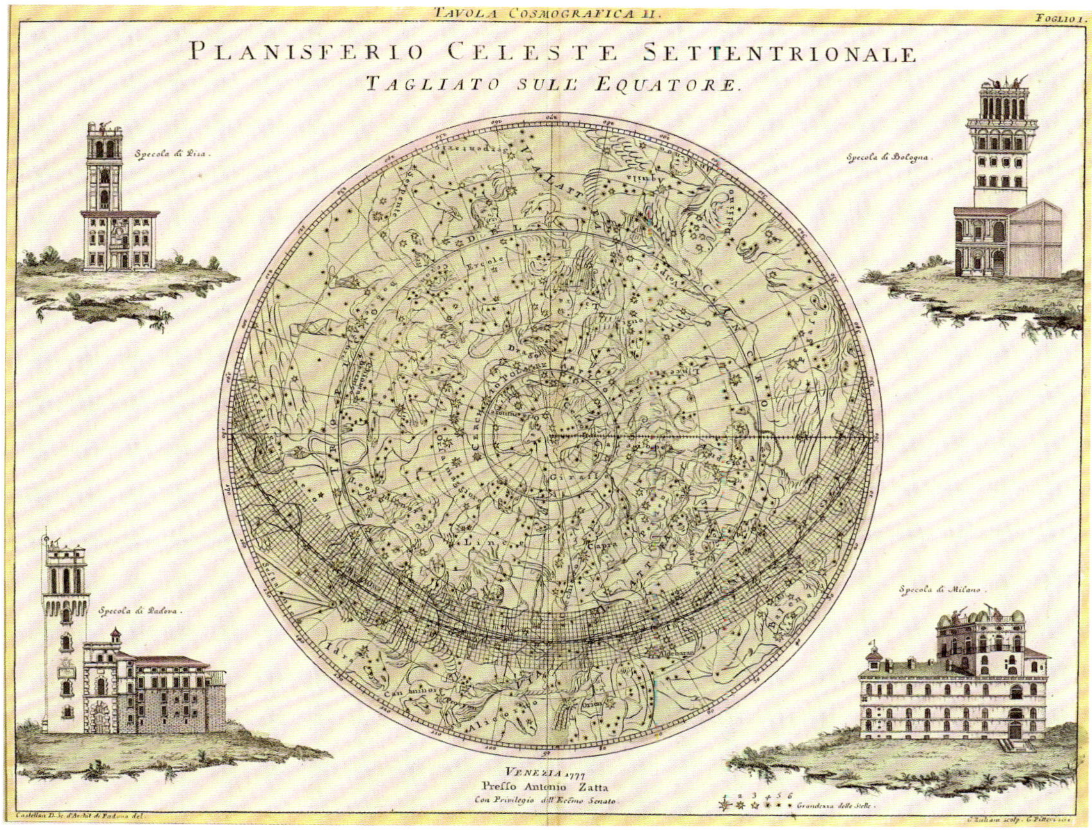

CAT. NO. 80 Antonio Zatta, *Planisferio Celeste Settentrionale Tagliato Sull' Equatore* from *Atlante Novissimo*, 1779

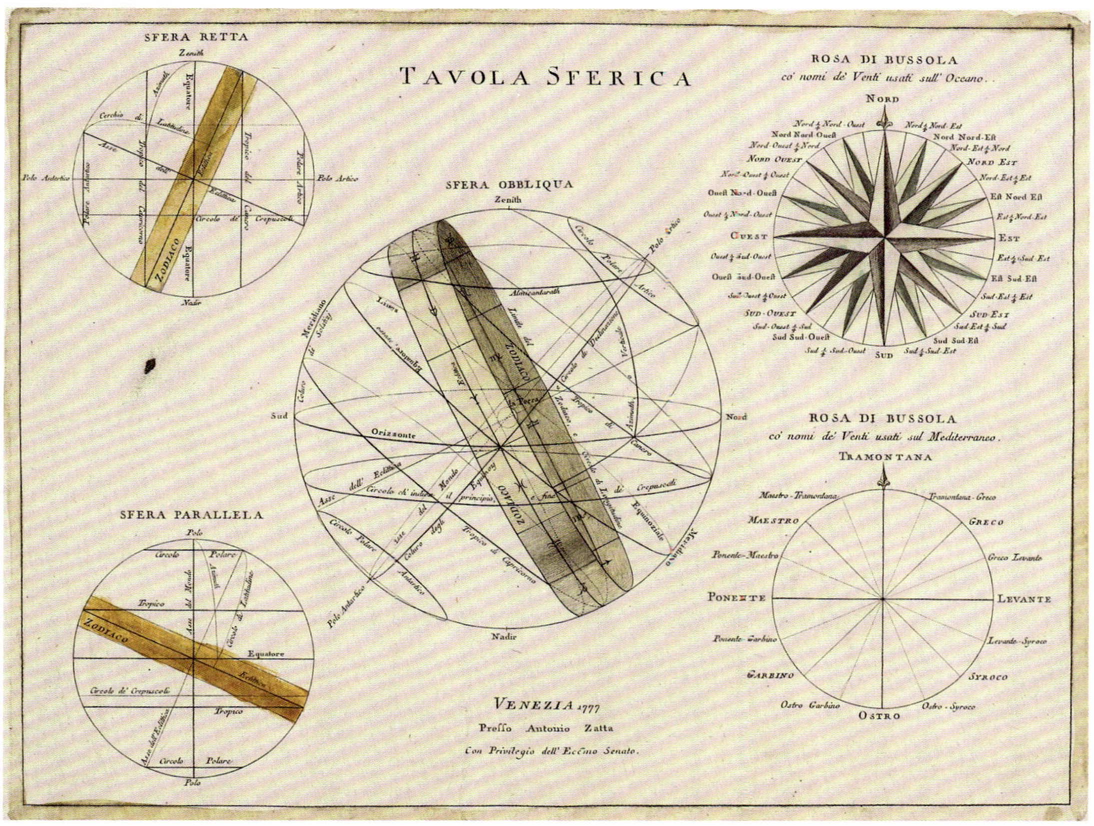

CAT. NO. 81 Antonio Zatta, *Tavola Sferica* from *Atlante Novissimo*, 1779

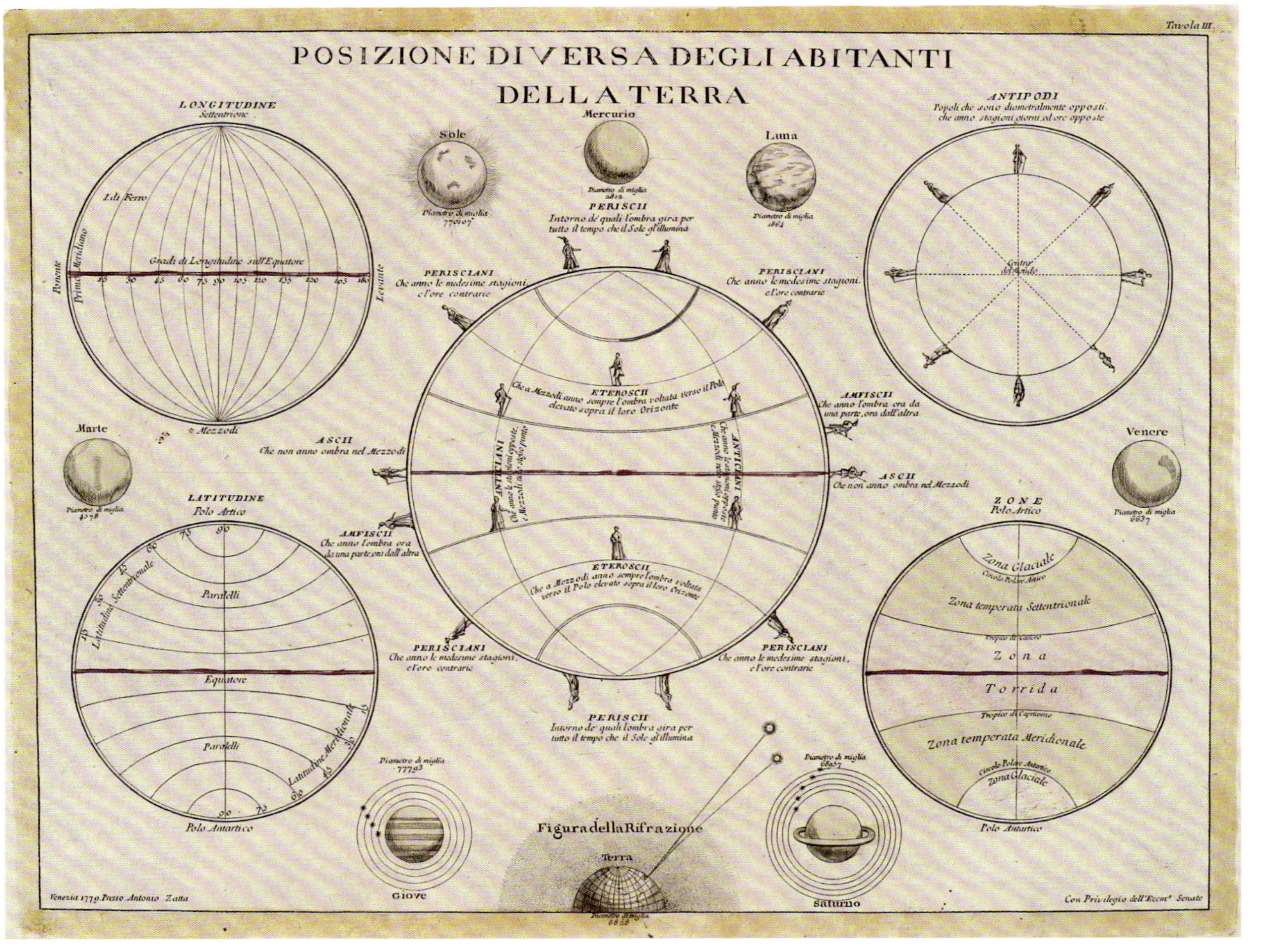

CAT. NO. 82 Antonio Zatta, *Posizione Diversa Degli Abitanti Della Terra*, 1779

CATALOGUE OF THE EXHIBITION

1. Frans Afferden (1653–1709)

Plani-Spherium Coeleste from *El Atlas abreviado*, Spain, 1696
Engraving 7½″ x 11⅞″

Frans Afferden was a Belgian cartographer who compiled *El Atlas abreviado*. The atlas was issued in several editions between 1696 and 1725 and contained a celestial planisphere, a flat reproduction of a hemisphere of stars. The constellation figures were derived from celestial inserts on the world map of the great seventeenth-century map publisher Joan Blaeu. The small size of the constellations prevents the inclusion of great detail, but the imaginative use of color in the Cetus (the whale or sea monster), for example, along with the lively gestures of the figures lend a dynamic quality to the work.

2. Johann Bayer (1572–1625)

Cassiopeia from *Uranometria*, Ulm, 1639
Engraving 14″ x 16″

3. Johann Bayer (1572–1625)

Cygnus from *Uranometria*, Ulm, 1639
Engraving 14″ x 16″

4. Johann Bayer (1572–1625)

Eridanus from *Uranometria*, Ulm, 1639
Engraving 14″ x 16″

5. Johann Bayer (1572–1625)

Orion from *Uranometria*, Ulm, 1639
Engraving 14″ x 16″

Originally published in Augsburg in 1603, Johann Bayer's *Uranometria* was reissued in Augsburg in 1624 from the same plates and many times after in Ulm beginning in 1639, as an atlas of constellation charts without accompanying text. An Augsburg lawyer, Bayer was also an amateur astronomer whose star catalogue contains more stars than any previous set of charts. Forty-eight maps embody the individual Ptolemaic constellations; an additional map shows the twelve new constellations of the southern skies as charted by the Dutch explorer Pietr Dirksz Keyser.

Bayer is best known for perfecting a lettering system denoting the brightness of stars that is still in use: Greek letters for brighter stars and Roman letters for fainter stars (such as the "Alpha" star in Orion's left shoulder). Art historians acclaim his work for the beauty of figure design and the quality of the engraving. *Cassiopeia*, who according to Milton, "strove / To set her beauty's praise above / The Sea-nymphs," is mounted on a severe, simply styled royal throne. The palm of victory that she holds is of doubtful meaning. The brightest star, at the left of her chair, is actually a supernova discovered by Tycho Brahe. Although seen from below (that is, geocentrically), with his claws clearly visible, delicately colored *Cygnus* appears to move gracefully across the northern skies. The fabled stream *Eridanus*, the longest constellation, is enhanced with rock formations, grass, and sedge in Bayer's version. The most vigorous of the male figures in the atlas (befitting his role as Hero and Warrior), *Orion* strides purposefully across the skies with eyes uplifted and brow furrowed in intense concentration. This chart is of particular interest because it shows signs of use: notations on specific observations have been made in tiny script at the right of the page. The four selections here can be paired with their Christianized counterparts by Bayer's contemporary and colleague, Julius Schiller (see CAT. NOS. 70–73).

6. John Bevis (1693–1771)

Corona Australis from *Atlas Celeste*, England, 1750
Engraving 12⅝″ x 14⅞″

7. John Bevis (1693–1771)

Crator and Corvus from *Atlas Celeste*, England, 1750
Engraving 12⅝″ x 14⅞″

8. John Bevis (1693–1771)

Sagitta from *Atlas Celeste*, England, 1786
Engraving 12½″ x 15″

9. John Bevis (1693–1771)

Ara from *Atlas Celeste*, England, 1786
Engraving 12½″ x 14⅞″

John Bevis was a British physician who turned to his avocation, astronomy, as a full-time occupation after practicing medicine for eight years. In 1738 he moved from London to Stoke Newington, where he built an observatory, and there he worked at compiling an atlas of the stars. The completed *Atlas Celeste* was patterned after Bayer's *Uranometria* with the addition of stars catalogued by Hevelius, Halley, Flamsteed, and Bevis himself corrected for the year 1746.

As in Bayer's example, there are forty-eight plates of the Ptolemaic constellations, one of the skies around the south equatorial pole, and two hemispheric maps. Although Bevis duplicates the Bayer iconography exactly, unlike Bayer he includes partial figures of all the constellations within the area of his chart, instead of only one at a time. The *Atlas Celeste* was nearly ready for publication when in 1750 the publisher John Neale went bankrupt. At that time, only a few copies of the printed charts had been circulated through subscription. Other copies eventually became available from Bevis's estate and were offered by booksellers in 1786. Today, fewer than twenty copies are known.

The image of the crown between the forelegs of Sagittarius was accepted in Ptolemy's time as one of the ancient forty-eight constellations. It has been variously explained through the centuries, as among other things: the Centaur's crown, and therefore the crown of Sagittarius (since centaurs were often represented wearing crowns), the crown that Bacchus placed in the sky in honor of his mother Semele, and as the bunch of arrows radiating from the hand of the archer (often imagined as a wheel). To the right side of the plate are portions of the tail of Scorpio and of the altar, shown in an inverted position as is appropriate for a southern constellation. This plate is inscribed to Francis Wollaston, an Anglican priest and amateur astronomer who himself published a comparative catalogue of the stars, a catalogue of circumpolar stars, and a collection of ten plates depicting the heavens as they appear to the naked eye.

Corvus was once a silver feathered raven that foretold Apollo that his mistress (Coronis) would be unfaithful. The outraged god cursed the bird and changed its feathers to "raven black." *Crator*, the heavenly goblet, also belonged to Apollo, but was shared with Dionysus, the ecstatic Greek god of wine.

A full view of *Ara* is also shown, a constellation named by Cicero that has generated many stories. One version describes it as the altar used to offer thanksgiving after the Great Flood, with its flames and sweet fragrances merging into the Milky Way (shown in gray tones behind it). Even more dramatically, it was said to be the altar built by the Cyclopes (weapon makers to the gods) upon which the first ever sacrifice was offered to Zeus. Its smoke merged with the Milky Way.

The Cyclopes story continues with *Sagitta* (the Arrow), the smallest of all the constellations. Hercules used it to end a vulture's torment of Prometheus, the Titan doomed in chains for eternal torture. It is also associated with the arrow of Cupid, and in its Christian version the lance that the Roman centurian Gaius Cassius used to end Christ's suffering on the cross. The inscription to Lord George Anson recognizes a famous British admiral whose four-year voyage around the world was one of the great achievements of naval heroism.

10. Richard Bloom (1641–1705)

Cosmography and Astrology from *The Gentleman's Recreation*, London, c. 1686
Engraving 14¾″ x 9⅜″

Richard Bloom produced a broad range of maps with engraving quality that spanned from the crude to the elegant. He is often referred to as a famous plagiarizer due to his unabashed (and often acknowledged) copying of Speed, Samson, and here De Sauvigny's *Tableaux Accomplis De Tous Les Arts* (1587). For *The Gentlemen's Recreation*, Bloom prepared several charts that described the various components of disciplines. In each case, the format resembles a modern flow chart for an organizational structure or a computer program. For the *Cosmography and Astrology* engraving here, astrology does not suggest its current pseudo-science meaning but rather is a name for practical astronomy.

11. Johann Elert Bode (1747–1826)

Title page from *Uranographia sive Astrorum Descriptio vinginti tabulis aeneis incisa ex recentissimis et absolutissimis Astronomorum observationibus*, Berlin, 1801
Engraving 24⅛″ x 34⅛″

12. Johann Elert Bode (1747–1826)

Virgo and Libra from *Uranographia sive Astrorum Descriptio vinginti tabulis aeneis incisa ex recentissimis et absolutissimis Astronomorum observationibus*, Berlin, 1801
Engraving 24⅛″ x 34⅛″

13. Johann Elert Bode (1747–1826)

Auriga and Lynx from *Uranographia sive Astrorum Descriptio vinginti tabulis aeneis incisa ex recentissimis et absolutissimis Astronomorum observationibus*, Berlin, 1801
Engraving 24⅛″ x 34⅛″

14. Johann Elert Bode (1747–1826)

Southern Pole Constellations from *Uranographia sive Astrorum Descriptio vinginti tabulis aeneis incisa ex recentissimis et absolutissimis Astronomorum observationibus*, Berlin, 1801
Engraving 24⅛″ x 34⅛″

Johann Elert Bode was a self-taught astronomer who became the director of the astronomical observatory of the Berlin Academy of Sciences, a post that he held for nearly forty years. During his lifetime he published a number of very popular astronomy texts and atlases, one of which was the *Uranographia*, a successor to the celestial atlases of Bayer, Hevelius, and Flamsteed. It was the last of its kind. Following its publication, constellation figures in star atlases would be schematically represented, if at all, because of the number of new celestial objects being continually charted.

The *Uranographia* contained twenty engraved plates, eighteen of which represented the major constellations with the two remaining representing the hemispheres. These plates accompanied Bode's star catalogue, the *Allgemeine Beschreibung und Nachweisung der Gestirne* that contained the stars of Flamsteed and twenty-eight other late seventeenth- and eighteenth-century astronomers, including Lalande, Lacaille, Messier, William Herschel, and Bode himself.

In this selection, the constellation *Virgo* is shown as an elegant, confident young woman worthy of a John Singer Sargent portrait. The constellation *Auriga* represents a charioteer or wagoner with a whip and horse bit in his right hand, a goat supported against his left shoulder, and two kids in the crook of his left arm. In some texts he is said to be Ericthonius, the son of Vulcan and Minerva, who, having inherited his father's lameness, found it necessary to invent an easy means of locomotion. His invention of the four-horse chariot earned him a place in the stars. Beside him is a representation of William Herschel's telescope, one of the instruments placed at a late date among the ancient constellations to glorify the astronomer and his occupation. Bode's plate of *Auriga* is particularly noteworthy because it shows a "pre-discovery" position of Uranus (see lower right corner in Taurus), a "star" recorded by Flamsteed in 1690, prior to Herschel's actual discovery of Uranus in the constellation Gemini in 1781. Such pre-discovery locations assist astronomers in determining the orbit of a newly discovered body.

The stars that form the *Southern Pole Constellations* are so packed with new and old constellations that the demise of the great star atlases becomes a necessity. There are robust renditions of animals, instruments, and a Native American. The wonderful depictions of constellations associated with birds make this Audubon-like swan song of celestial cartography a particularly exciting image.

15. Elijah Hinsdale Burritt (1794–1838)

The Visible Heavens in July, August and September and *The Visible Heavens in April, May and June* from *The Atlas Designed to Illustrate the Geography of the Heavens*, New York, 1835
Engravings 16½″ x 14½″ each

Connecticut-born Elijah Burritt, educator and journalist, designed his celestial atlas as an inexpensive alternative to a celestial globe, heretofore the principal educational aid for teaching elementary astronomy. A textbook, *The Geography of the Heavens*, accompanied the atlas. In place of globe gores, Burritt introduced four maps designating star locations by season. The constellation figures are obvious copies from Alexander Jamieson's atlas published in London in 1822, *A Celestial Atlas*. Compare the left page images of Hercules and Ophiuchus to Jamieson (CAT. NOS. 56–57), themselves fashioned after Bode (CAT. NOS. 11–14).

The combined text and atlas sold originally for $1.25 (equivalent to about $26 in 2005); this was more than the cost of other introductory astronomy texts, but far less than the price of a globe. Unlike other textbooks, Burritt's *Atlas* and *Geography* emphasized stellar astronomy and encouraged students to perform their own observations. Enthusiastic reviews greeted the publications. A professor of natural philosophy at Yale University praised Burritt for helping to "diffuse a popular knowledge of astronomy." Burritt died prematurely of yellow fever at the age of forty-four, but his atlas continued to sell, with 300,000 copies in print by 1876. For more information, see Peggy Aldrich Kidwell, "Elijah Burritt and the 'Geography of the Heavens,'" *Sky & Telescope* (January 1985): 26–28.

16. Claude Buy de Mornas (eighteenth century)

Systèmes de Ptolomée et de Ticho Brahé from *Atlas Méthodique et Elémentaire de Géographie et d'Historie*, Paris, 1761
Engraving 15⅞″ x 21⅞″

17. Claude Buy de Mornas (eighteenth century)

Système de Descartes from *Atlas Méthodique et Elémentaire de Géographie et d'Historie*, Paris, 1761
Engraving 15⅞″ x 21⅞″

Buy de Mornas was a prominent French mapmaker of the mid-eighteenth century. His atlases typically had double page plates with text alongside, similar to the Buchor/Carey and Lea style of cartography. Each image contains an internally printed frame-like border that is highly decorative and helps to focus the reader's attention to the material within. In *Systèmes de Ptolomée et de Ticho Brahé* the geocentric cosmology of Ptolemy and the combined geocentric-heliocentric system of Brahe are depicted.

The portrayal of the cosmological theory of René Descartes (1596–1650) is particularly interesting in *Système de Descartes*. While a supporter of the new heliocentric system of Copernicus, Descartes withheld publication of his *Le Monde* (1634) once he heard about Galileo's trial in 1633. Superficially a re-statement of Copernicus, the engraving of Buy de Mornas actually updates what Descartes offered in his *Principia Philosophiae* (1644). Jupiter not only has the four moons discovered by Galileo (1610), but now Saturn is shown with moons discovered by Huygens (1655) and by Cassini (1671, 1672, 1684), all well after Descartes' death. Most significant is the portrayal of Descartes' view that stars are not all at the same distance from our solar system. The decorative swirls are not adornments, but the key idea of Descartes' "Vortex Theory" of cosmology: that planetary motions are driven by immense whirlpools of invisible matter emanating from the Sun and other stars. The idea was to be supplanted soon by Newton's gravitational action-at-a-distance.

18. Giovanni Maria Cassini (c. 1790)

Celestial Globe Gores: Cassiopeia, Perseo, Orione from *Globo Celeste*, Rome, 1792
Engraving 19¾″ x 13⅜″

19. Giovanni Maria Cassini (c. 1790)

Celestial Globe Gores: Gemelli, Cancro, Leone from *Globo Celeste*, Rome, 1792
Engraving 19¾″ x 13⅜″

20. Giovanni Maria Cassini (c. 1790)

Celestial Globe Gores: Vergine, Boote, Ercole from *Globo Celeste*, Rome, 1792
Engraving 19¾″ x 13⅜″

21. Giovanni Maria Cassini (c. 1790)

Celestial Globe Gores: Sagitarrio, Capricorne, Aquario from *Globo Celeste*, Rome, 1792
Engraving 19¾″ x 13⅜″

22. Giovanni Maria Cassini (c. 1790)

Tavola Sferica, Rome, 1788
Engraving 13½″ x 18⅝″

Giovanni Maria Cassini (not to be confused with the French Cassini family) was an Italian globe maker, engraver, and geographer of distinction. These celestial globe gores (sections of a global map) are based on the astronomical observations of Flamsteed and Lacaille. Stars vary in size and thereby represent different magnitudes. The Ptolemaic constellations are interspersed with modern figures that have been added in dotted lines, e.g., the Unicorn (in CAT. NO. 21), a constellation that came into use only in the third quarter of the sixteenth century. Delicately drawn and softly colored, the constellation figures give a lyrical version of the star myths. Among the post-classical constellations represented are the American Indian (Indus) in the first set of gores, the Unicorn in the third set of gores, and astronomical instruments shown as dotted lines in several gores. The constellations have been shaded to show depth and roundness but appear somewhat awkward in their proportions. The light pastel hues used to color the figures echo the palette of the rococo era earlier in the eighteenth century.

Cassini's *Tavola Sferica* was in all probability copied from the work of Antonio Zatta (see CAT. NO. 81). In this plan, the Earth is shown at the center of the sphere for the purpose of locating objects in the sky. The circles that surround the Earth are meant to facilitate celestial observations by marking off the Universe into sections. The "Sphera retta" and the "Sphera parallela" show cross sections of this central sphere useful to an observer standing at the equator and at some northern latitude. The two *Rose dei bussole* that adorn the right side of the engraving list the names of the winds of the ocean and the Mediterranean.

The lovely cartouches that adorn the title, the printing information, and the date on this engraving add a charming and lively touch to an otherwise austere plan. The baroque elements of flying creatures and a personification of the wind are here subordinated to the astronomical information imparted by the plans. The Age of Enlightenment has relegated formerly important decorative elements to a minor role.

23. Andreas Cellarius (eighteenth century)

Situs Terrae Circulus Coelestibus from *Atlas Coelestis seu Harmonia Macrocosmica* (1st edition, 2nd printing), Amsterdam, 1662
Engraving 17⅛″ x 20¼″

24. Andreas Cellarius (eighteenth century)

Typus Selenographicus from *Atlas Coelestis seu Harmonia Macrocosmica* (1st edition, 2nd printing), Amsterdam, 1662
Engraving 17⅛″ x 20⅜″

25. Andreas Cellarius (eighteenth century)

Planisphaerium Braheum from *Atlas Coelestis seu Harmonia Macrocosmica* (1st edition, 2nd printing), Amsterdam, 1662
Engraving 16¾″ x 20⅝″

26. Andreas Cellarius (eighteenth century)

Planisphaerium Braheum from *Atlas Coelestis seu Harmonia Macrocosmica* (2nd edition), Amsterdam, 1708 (Valk and Schenck)
Engraving 16⅝″ x 20¼″

27. Andreas Cellarius (eighteenth century)

Tychonis Brahe Calculus Planetarium, from *Atlas Coelestis seu Harmonia Macrocosmica* (1st edition, 1st printing), Amsterdam, 1660
Engraving 16⅞″ x 19⅞″

28. Andreas Cellarius (eighteenth century)

Theoria de Veneris et Mercurii from *Atlas Coelestis seu Harmonia Macrocosmica* (2nd edition), Amsterdam, 1708 (Valk and Schenck)
Engraving 17″ x 20¼″

29. Andreas Cellarius (eighteenth century)

Hypothesis Ptolemaica from *Atlas Coelestis seu Harmonia Macrocosmica* (2nd edition), Amsterdam, 1708 (Valk and Schenck)
Engraving 16⅞″ x 19⅞″

30. Andreas Cellarius (eighteenth century)

Hypothesis Ptolemaica from *Atlas Coelestis seu Harmonia Macrocosmica* (1st edition, 2nd printing), Amsterdam, 1662
Engraving 16⅞″ x 19⅞″

31. Andreas Cellarius (eighteenth century)

Hypothesis Ptolemaica from *Atlas Coelestis seu Harmonia Macrocosmica* (1st edition, 1st printing), Amsterdam, 1660
Engraving 16⅞″ x 19⅞″

The *Harmonia Marcrocosmica* of Cellarius is a beautiful bound volume containing numerous charts and a wealth of astronomical information. After a lengthy prologue that places the ideas of ancient and modern astronomers in historical sequence and discusses the physical nature of the Universe, the following chapters outline in detail the theories of Ptolemy, Copernicus, Tycho Brahe, and Aratus. The second half of the volume discusses more general matters such as the magnitudes of the stars, lunar and solar theory, the nature of the planets, the hemispheres, and the zodiac. Each chapter is beautifully illustrated with colorful plates, all of which are individual in nature. Their ornament is varied and exquisitely executed, showing the care that went into the design of each engraving.

Situs Terrae Circulus Coelestibus is a splendid "schematic" of the celestial machinery, where Cellarius depicts the projections of the Earth's poles and equator upon the celestial sphere. From the frame of reference of a fixed Earth, the celestial bodies are seen to move along various circular paths in the sky. The path of the Sun is the Ecliptic, inclined 23½ degrees with respect to the celestial equator. The 12-degree band of the sky that spans the Ecliptic and contains the paths of all the planets is called the Zodiac; it is divided into twelve regions or signs named in antiquity.

The central Earth is given a special reality by the mapping of actual geography, at least as known in 1660. Note, for example, the fairly accurate depiction of India and Southeast Asia, as well as the then commonly held view that California was an island of uncertain structure.

Typus Selenographicus shows how the relative positions of the Sun and Moon result in the phases of the Moon as seen from Earth. This explanation dates to Aristotle, who argued that the Moon shines by reflected sunlight, and thus that lunar phases result from changed viewing perspectives. The plan is geocentric; but since the Moon does indeed orbit the Earth, and the Sun is far beyond the Moon's orbit, this is a perfectly functional working model. In the lower left of the chart is an elaborate plan of the Moon's phases, listing thirty-six separate aspects that might be seen during the course of a month. The plan in the lower right corner shows in a more detailed, scientific manner the fact that half of the Moon is illuminated by sunlight, and how the viewing geometry determines the resulting phase seen from Earth.

This plate is a classic example of the interplay between art and science in the seventeenth century. The rays of the Sun form a colorful backdrop to the diagram; putti gaily hold the titled banners and astronomical measuring instruments, all under the benevolent eyes of the Sun. The Dutch engraver J. van Loon here celebrates science with the exuberance and beauty characteristic of the age in which he lived.

Planisphaerium Braheum depicts the interesting plan of Tycho Brahe, the eccentric Danish nobleman and astronomer of the sixteenth century. Tycho's plan was a compromise between those of Ptolemy and Copernicus in that it made the Sun revolve around a stationary central Earth while all the other planets revolved around the Sun. Tycho acknowledged parts of the Copernican doctrine but refused to believe that the solar system was heliocentric, partly because of his respect for Scripture but also because he had not observed any stellar parallax or the apparent annual displacement of nearby stars, the necessary consequence of the Earth's motion around the Sun. His geoheliocentric system gained acceptance in some quarters in part because it did not bring him into conflict with theologians. In his observations of the motion of comets through space, Tycho had come to reject the Aristotelian theory that held that the planets were attached to spheres and rotated with them. With his reappraisal of the Universe, Tycho helped prepare the way for the eventual universal acceptance of the system of Copernicus.

This beautifully colored and designed plate presents Tycho's ideas with unabashed charm. In a festively colored sky, baroque cherubim talk among themselves and admire Tycho's plan. In a more serious vein, the vignettes of astronomers making observations and teaching their young students what they have deduced from the heavens salute learning and knowledge for its own sake. The search for an explanation to the Universe, the desire to know and understand the world is here honored as a noble endeavor. Science and art are once again seen to complement each other, and their interdependence successfully carries on the humanist values awakened in the Renaissance.

In *Tychonis Brahe Calculus Planetarium*, another plate devoted to the system of Tycho Brahe, Cellarius presents more details. A tiny Earth is at the very center, with planets circling around the Sun, itself in motion about the Earth.

Theoria de Veneris et Mercurii depicts Mercury and Venus, the planets closest to the Sun, and therefore ones that never appear to be very distant from it. Consequently, they are visible in the hours near sunrise and sunset (and never at midnight, e.g., near the full Moon). This "following the Sun" aspect enforces the geocentric notion, and here Cellarius gives all the wonderfully complicated epicyclic details when they are closest to the Sun (perihelion) and farthest (aphelion). This plate from the Valk and Schenck edition is austere in the use of color, resulting in a dramatic focus upon the central elements.

Hypothesis Ptolemaica illustrates the theory of eccentrics and epicycles used in the second century A.D. by the Egyptian astronomer Claudius Ptolemaeus. Basing his work on that of earlier Greek astronomers and Babylonian observations, Ptolemy developed a geocentric system of the Universe that was so satisfactory that it remained unchallenged in the West for fourteen hundred years. Though earlier Greek astronomers (e.g., Hipparchus) had developed geocentric systems using eccentrics and epicycles for the Sun, none had produced a model with the five planets then known.

The motion of the Sun, Moon, and planets in Ptolemy's system was not uniform rotation in concentric circles around the Earth. Instead, movement was determined by a more complicated set of circular orbits that did not have the Earth precisely at the center. The paths of the planets were therefore eccentric or off-center (as illustrated in the diagram in the lower left corner of Cellarius' chart). In this way, a body moving around the Earth would at times be nearer and at other times farther from the Earth. Moreover, the bodies rotating about the Earth also moved in an epicycle: that is, they each rotated in a circle whose center moved along the circumference of the eccentric orbit (as shown in the lower right diagram of the chart). This type of eccentric and epicyclic movement was necessary in a geocentric system in order to explain the seeming irregularity of the motion of the Sun, Moon, and planets and to predict in a reliable manner the lunar and solar eclipses. It is a tribute to the genius of Ptolemy that such a complex system could be devised. It was not so much intended to describe reality as to provide a geometrical scheme for computing positions of the planets. Today we know that planets move in elliptical paths and at speeds that are not constant; this type of explanation would not have appealed to the ancient Greek mind, which sought a system of the cosmos using only uniform circular motions.

The ornate decorations filling the spaces between diagrams in this work take the form of shells, unusual faces, and what appear to be seahorses. In their lively baroque manner, they add vitality and expressive quality to simply designed diagrams.

32. Vincenzo Maria Coronelli (1650–1718)

Hercules (Celestial Globe Gore) from *Oribis Coelestis Typus*, Paris, (twentieth-century re-strike from original plates dated 1693)
Engraving 26½″ x 12″

33. Vincenzo Maria Coronelli (1650–1718)

Serpens (Celestial Globe Gore) from *Oribis Coelestis Typus*, Paris, (twentieth-century re-strike from original plates dated 1693)
Engraving 26½″ x 12″

34. Vincenzo Maria Coronelli (1650–1718)

Gemini (Celestial Globe Gore) from *Oribis Coelestis Typus*, Paris, (twentieth-century re-strike from original plates dated 1693)
Engraving 26½″ x 12″

35. Vincenzo Maria Coronelli (1650–1718)

Canis Major (Celestial Globe Gore) from *Oribis Coelestis Typus*, Paris, (twentieth-century re-strike from original plates dated 1693)
Engraving 26½″ x 12″

36. Vincenzo Maria Coronelli (1650–1718)

Set of Gores for a Nine-inch Celestial Globe, Venice, 1700
Engraving 9⅝″ x 18¾″

37. Vincenzo Maria Coronelli (1650–1718)

Primi Elementi, Venice, 1690
Engraving 18⅜″ x 24⅜″

Grandeur of conception, largeness of scale, and magnificence of design characterize the globes and gores of Vincenzo Maria Coronelli, the great seventeenth-century Venetian cartographer. More than four hundred maps and many globes credited to him were made at the Franciscan convent outside of Venice, where Coronelli lived with the members of his religious order. Through his achievement, the seat of excellence in cartographic publishing passed from Amsterdam to Venice at the end of the seventeenth century. Ornate globes, unlikely products of a modest Franciscan workshop, were made for the most discriminating European royalty. With a diameter of fifteen feet, the celestial globe Coronelli made for Louis XIV of France was the largest in existence until the twentieth century.

The gores from the 1693 edition of Coronelli's celestial globe gores probably represent a slightly modified version of the Venice globe gores issued in 1688. The Musée du Louvre, which owns the original plates from which the gores were printed in 1693, issued a small number of sets of re-strikes in the early twentieth century, of which the four gores here are examples.

Energetically drawn constellation figures populate Coronelli's gores. Richly detailed, they project the illusionism of the Baroque. Taut flesh and rippling muscles are flourishingly depicted in *Hercules*. Below, the northern hemisphere globe section is continued into the southern hemisphere with the constellation *Serpens*.

For the twin boys (Gemini) and the small dog (Canis Minor), Coronelli offers an emotional appeal unequaled in celestial cartography. Castor and Pollux, the Heavenly Brothers, were the offspring of Zeus and Tyndareus. They traveled with the Argonauts and are famous for calming the rough seas. The southern hemisphere globe section completes Monoceros (the Unicorn, shown in *Gemini*, CAT. NO. 34), a constellation not based upon legend. Below is the Great Dog in full splendor. Sirius, the brightest star in the night sky, is the "Dog Star," shown in the nose. Its appearance in July and August gives name to the "dog days of summer" in the northern hemisphere. Together, these four gores represent the summation of the artistic achievement in constellation design in the High Baroque period.

The set of twelve globe sections in *Set of Gores for a Nine-inch Celestial Globe* (CAT. NO. 36) would have been cut out and pasted to a six-inch sphere to be displayed with a matching geographical globe. The miniature constellation figures leave little opportunity for the full talents of Coronelli to be shown. Compare the Hercules-Serpens gore (far left) with the same part of the celestial sphere depicted in the gores for a four-foot globe displayed in *Hercules* and *Serpens* (CAT. NOS. 32 & 33).

The sumptuous images used in the first few constellation portrayals by Coronelli are in remarkable contrast to the highly functional descriptions of coordinate systems and geometrical ordering of terrestrial and celestial spheres in *Primi Elementi*. In the upper left, Coronelli proudly displays his title of "Cosmographer to the Serene Republic of Venice," and to the right a dedication to a relative of the meritorious Senator Belegno. Of special interest are the tables at the bottom that give various values found for the angular separation of the tropics from the equator, or the "obliquity of the ecliptic": the cause of season due the angular tilt of the Earth's axis of rotation with respect to the plane containing Earth and Sun. From the ancient astronomers of antiquity (e.g., Ptolemy in 140 A.D.) to the most recent determinations by D. Cassini in 1687 (just a few years before the engraving of this chart), the value of approximately 23 degrees is shown to be well known.

38. Nicolas de Fer (1646–1720) with H. van Loon (1649–?), engraver

Four Systems of Cosmology from *Atlas Historique*, Paris, 1705
Engraving 9⅜″ x 13⅛″

39. Nicolas de Fer (1646–1720) with H. van Loon (1649–?), engraver

Planispheres Celeste, by de la Hire from *L'Atlas Curieux*, Paris, 1702
Engraving 9⅜″ x 13¼″

Nicolas de Fer was a French engraver and geographer who, among other accomplishments, published several maps of France. His maps, however, were known for their ingenious ornament rather than their geographical exactitude. The engraver of this print, Hermanus van Loon, was a Flemish artist residing in Paris who worked with de Fer on more than one of his projects.

The systems of Ptolemy, Copernicus, Descartes, and Tycho Brahe are juxtaposed here for comparison. Differences such as the spacing between planets and the position of the firmament of the stars as well as the differences between geocentric, heliocentric, and geo-heliocentric systems are made explicit. It is interesting that de Fer has taken the liberty of updating both Copernicus's and Tycho's plans by showing discoveries that were not made until later: the four moons of Jupiter (discovered by Galileo in 1610) and five of the moons of Saturn, which were discovered in 1684. The clouds that surround each system indicate the still-incomplete knowledge of the extent of the Universe. This engraving is relatively unembellished for its time. Only pale pastel coloring and rather fanciful clouds enliven its pedagogical nature.

A more impressive display of engraving appears in the plate of twin planispheres (flat, circular renditions of a hemisphere of stars). Of note is the choice of a coordinate system not based upon the usual celestial poles and equator, but one reckoned with respect to the Earth's orbit and its pole (the ecliptic system). In this scheme, Ursa Major (Big Bear) is in the lower left with its pointer stars towards Polaris not in the center of the system. In each corner, the latest telescopic discoveries of the French Cassini family of astronomers are shown: remarkable solar protuberances, a map of the Moon, and early attempts to identify features on Mars, Venus, and Saturn.

40. Denis Diderot (1713–1784)

Hemisphere Boreal, from *Encyclopédie ou Dictionnaire Raisonné des Sciences des Arts et des Métiers*, Paris, 1751–1780 (illustrations published 1762–1777)
Engravings 13¾″ x 8¾″

41. Denis Diderot (1713–1784)

Hemisphere Austral from *Encyclopédie ou Dictionnaire Raisonné des Sciences des Arts et des Métiers*, Paris, 1751–1780 (illustrations published 1762–1777)
Engravings 13¾″ x 8¾″

These two hemispheres once formed part of a volume of plates from Diderot's *Encyclopédie*. They would have helped to illustrate the section on astronomy written by the mathematician and philosopher Jean le Rond d'Alembert (1717–1783), the author of the mathematical sections of the *Encyclopédie*. In his essay, d'Alembert traces the history of astronomy from ancient times up to his own day, paying particular attention to names and nationalities of astronomers while only briefly noting the changing conception of the Universe through the centuries. D'Alembert himself advocated Newton's theories of gravitation in explaining celestial phenomena, as was common for his time.

The eleven volumes of illustrations for the *Encyclopédie* were published from 1762 to 1777 amid angry charges of their having been taken from engravings prepared for the Academy of Sciences. A project of the Academy since its founding in 1666 had been to maintain a scientific surveillance over French industry. In 1757, its director, the naturalist and metallurgist Réaumur, had commissioned a large number of plates, with the intention of systematically recording French industrial life and methods. Diderot must have found, bought, or bribed his way to Réaumur's plates during the early stages of publishing the *Encyclopédie*, but it should be noted that copyright did not exist in the eighteenth century and the title to artistic and literary property was not legislated.

The *Hémisphere Boréal* depicts the constellations of the northern hemisphere with their French names. Several of the constellations instituted by Hevelius, one by Plancius (the giraffe), and one by Le Monnier (the reindeer) have been added in dotted lines to those of Ptolemy. The dogs of Boötes probably date from as early as the second quarter of the sixteenth century, while the unicorn may date to the third quarter of that century. In the corners of this plate, the stars of Cygnus, Cassiopeia, Ursa Major, Ursa Minor, and the head of Andromeda are represented without their respective mythical configurations.

The *Hémisphere Austral* is less densely populated than its northern counterpart because of the still incomplete knowledge of the stars of the southern hemisphere. In the corners of this plate are the star formations of the crane, Scorpio, Rigel, the foot of the centaur, and the nearby small cross. The figures of these two hemispheres are sketchily drawn after earlier baroque examples.

42. Johann Gabriel Doppelmayer (1671?–1750) and Johann Baptist Homann (1644–1724)

Motus in Coelo Spirales, Nuremberg, c. 1720
Engraving 20½″ x 23¾″

43. Johann Gabriel Doppelmayer (1671?–1750) and Johann Baptist Homann (1644–1724)

Phaenomena, Nuremberg, c. 1720
Engraving 20½″ x 23¾″

44. Johann Gabriel Doppelmayer (1671?–1750)

Phaenomena in Planetis Primariis, Nuremberg, c. 1720
Engraving 20″ x 23¼″

Johann Gabriel Doppelmayer, professor of mathematics in Nuremberg, and Johann Baptist Homann, founder of the great cartographic publishing house in Nuremberg, collaborated for over twenty years in producing astronomical charts and a celestial atlas. Doppelmayer's expertise in natural philosophy (today, the independent fields of astronomy, geography, mathematics, physics, and the history of science) merged quite effectively with Homann's reputation as a cartographer to produce a series of charts that contained an extraordinarily large amount of astronomical information. Accordingly, the splendor seen in Cellarius (CAT. NOS. 23–31) is reduced to a much more limited portion of each chart by Doppelmayer and Homann.

Phaenomena includes in a series of figures a description of how the seasons occur on Earth, followed by seven cosmological systems. In contrast to the three well known solar system theories (those of Ptolemy, Tycho, and Copernicus), Doppelmayer begins in the upper left with Ptolemaicum et Alphonsinum, and then moves on to Plutarchianum, with Platonicum at the upper right, and below it Porphyrianum. Along the bottom are the systems of Jacob Coccaei, Gilbert Angli, and Sebastiani Clerici. With the exception of Ptolemy these names are not part of the popular history of astronomy. The upper four systems are geocentric, while the lower three are partially heliocentric. The system by Angli is quasi-heliocentric, somewhat like Tycho's, and shows the stars occupying an extended region beyond the planets (not in a narrow shell, as in the other systems). In De Fer's *Four Systems of Cosmology* (CAT. NO. 38), Descartes' system goes beyond this to suggest that the domain of the stars does not have a clear outer boundary.

In the center of the image, considerable space is devoted to illustrating how the seasons occur. Using the heliocentric model that was well-established by this date, Doppelmayer shows the tilt of the Earth's axis of rotation as being away from the Sun at the top of the figure (and hence depicting winter solstice in the northern hemisphere); the tilt is toward the Sun at the bottom (summer solstice), with vernal and autumnal equinox conditions at left and right, respectively. The terrestrial geography shows clearly recognizable continents. The telescopic instruments tied to the central cartouche emphasize the increasing importance of precise observing methods to the history of astronomical thought.

In *Motus in Coelo Spirales*, Doppelmayer reviews how Tycho Brahe's system would account for the motions of Venus and Mercury. As described in more detail under Cellarius (CAT. NOS. 25–28), Tycho's was a compromise system in that the planets moved about the Sun, but the Sun and planets then revolved about the Earth. As with most compromises on important issues, it enjoyed little success; yet it is reproduced in virtually every account of the history of astronomy. As a computational device (and surely that is all Tycho meant it to be), the "true system" of Keplerian motion can be mimicked with some success using intricate combinations of circular, geocentric motions. The varying speeds and distances between the Earth and the two "inferior planets" (i.e., closer to the Sun) are handled by epicycles (motion along small circles) tied to larger circular motions, as illustrated in the upper right corner. The end result is that Mercury and Venus follow spiral patterns as seen from a fixed central Earth. In the cases computed here, the track of Venus (between 1 January 1712 and 27 April 1715) is shown in a red spiral; the track of Mercury (between 1 January 1712 and 28 December 1712) is shown in green. The use of epicycles to account for the so-called "retrograde motion loops" that occur when the Earth passes or is passed by a planet safeguarded the geocentric view for more than fourteen hundred years.

Embellishments normally found in the corners and sides of such charts are here replaced by specific examples and paragraphs of Latin text. In the center, personifications of Venus and Mercury swing from their respective spirals; each wears a star symbolic of the morning and evening appearances of these planets. Above the central Earth sits Apollo, gently pushing Venus along her path, while Mercury swings backwards in the retrograde part of his journey.

In *Phaenomena in Planetis Primariis*, a companion plate to *Phaenomena*, the scientific yield of telescopic observations of the planets bursts forth from all portions of the image. The central portions show the heliocentric system with inner planets (Mercury, Venus and Earth) on top, a zoomed-in version of the full solar system with outer planets Mars, Jupiter and Saturn shown below. Along the borders are portrayals of surface markings of the planets Venus, Mars, Jupiter and Saturn, showing how more and more details emerged as successive observers used improved telescopes, or more vivid imagination as in the case of Venus where, in reality, clouds obscure all details when viewed visually.

45. Charles-François Dupuis (1742–1809)

Jardin d'Ormusd, ou les six milles de Dieu, et les six Milles du Diable, ou l'Empire d'Ahriman from *l'Origine de tous les cultes ou religion universelle*, Paris (?), 1795
Engraving 9½″ x 8¾″

46. Charles-François Dupuis (1742–1809)

Planisphere astrologique de style Egyptian from *l'Origine de tous les cultes ou religion universelle*, Paris (?), 1795
Engraving 9½″ x 8¾″

In the early years following the French Revolution, Dupuis published his book on the "cults," perhaps best described as "beliefs," of Universal Religion. It contained twenty-two engravings, two of which are shown here. Later editions with expanded explanations followed in 1822 and 1835. In this print showing a fragment of Egyptian renditions of the familiar constellations of the zodiac, sub-divisions are given that divide the monthly transitions between the twelve zodiac signs into thirty-six segments, each with its own symbol. These 10-day segments are called Decans or Decanates, and are used for "high precision" astrological scenarios. Note that the center of the system is at the ecliptic pole (with the constellation Drago or Dragon), not Ursa Major of the equatorial system, because it is the Sun's passage through the sky (along the "ecliptic") that is of relevance to astrology.

In *Jardin d'Ormusd*, a far more complex plate, Dupuis uses the twelve constellations, again centered at the ecliptic pole, to summarize a "cult" that deals with an epoch of happiness when Man lived in Ormusd's Garden (shown in the top half) versus an equal period of darkness in the Empire of the Devil, or Ahriman. Along the upper rim, productive periods, such as harvesting grapes are identified with Virgo, and long days with Gemini. During the dark periods, wind and snow are associated with Sagittarius (bottom). In both plates, the constellation images are not finely drawn. There is no possible use intended here for locating stars, but rather a purely astrological and colorful set of sketches intended for popular consumption.

47. Jean Fortin (1750–1831)

Le Capricorne, et Le Verseau from *Atlas Celeste of John Flamsteed (1646–1719)*, Paris, 1776
Engraving 7⅜″ x 9⅜″

48. Jean Fortin (1750–1831)

L' Eridan, Orion, et Le Lievre, from *Atlas Celeste of John Flamsteed (1646–1719)*, Paris, 1776
Engraving 7¾″ x 9⅝″

49. Jean Fortin (1750–1831)

Le Bouvier, Les Lévriers, La Chevelute de Berenice from *Atlas Celeste of John Flamsteed (1646–1719)*, Paris, 1776
Engraving 7¾″ x 9⅜″

Jean Fortin was a French craftsman who made globes and spheres for the French king and the royal family. He is also known for his invention of the transportable mercury barometer. In 1776 he published the second edition of *Atlas Celeste* by the British astronomer John Flamsteed.

Le Capricorne, et le Verseau highlights the zodiacal constellations of Capricorn and Aquarius. Capricorn, or the goat-fish, is a very ancient sign that originated millennia ago in the region of the Tigris and Euphrates rivers. At times it has been represented as a complete goat-like animal, but its more usual form is a goat head and foreparts with the tail of a fish, as is seen in Fortin's engraving. The ancient Greek Platonists believed that when souls were released from the body, they ascended to Heaven through the stars of Capricorn and thus the sign was known as the "Gate of the Gods." Aquarius, the water carrier, so called because the Sun passes through this constellation during the rainy season, is usually represented pouring water from a bucket or urn. The fish depicted near Aquarius relate to his connection with water.

Fortin's constellations, as based on Flamsteed's work, are brightly colored and classicizing figures that reflect the contemporary interest in the ancient past. In the Capricorn engraving, the background has been colored a light blue, which offsets the figures it surrounds. The stars and highlighted features of the figures have also been delineated with a metallic gold paint.

In *Le Bouvier, Les Levriers, La Chevelute de Berenice*, Boötes (Le Bouvier) is represented as a herdsman with upraised arms in pursuit of the Great Bear. He holds a herdsman's staff, a scythe, and the leashes of his two hounds. In other constellation drawings, Boötes is known as Arcas, the son of Zeus and Callisto, a nymph in the train of the huntress Artemis who had been turned into a she-bear by that goddess (or, in some versions, by Zeus or Hera). Boötes thus pursues his own mother through the skies.

50. Isaac Frost (nineteenth century) with W. P. Clubb & Son, engravers

The Newtonian System of the Universe, The Newtonian Sys-

tem, System According to the Holy Scriptures, System According to the Holy Scriptures, from *Two Systems of Astronomy*, a Muggeltonian publication with oil printing by G. Baxter (1804–1867), London, 1846
Engravings 8½″ x 4¼″ each

The seventeenth-century religious sect known as the Muggletonians (co-founded by John Reeve and Lodowick Muggleton in England in 1651) held to a fundamentalist view in an era of great intellectual advancement in Natural Philosophy. Rejecting new observations in their astronomy, non-biblical causes in their physics, and the idea that mathematical calculations could characterize motions, the Muggletonians enjoyed certain popularity among political skeptics, religious zealots and anti-intellectuals in a turbulent Britain. In 1846, well after the widespread acceptance of a vast heliocentric planetary system, Isaac Frost published *Two Systems of Astronomy* in London giving the Victorian Muggletonians beautifully illustrated distinctions between the *System According to Holy Scripture* and *The Newtonian System of the Universe*. Included here are two examples from each system as portrayed by the new "oil color printing" method developed by George Baxter (1804–1867). The Muggletonian views are geocentric, with Moon, planets and stars within several miles of the Earth (and therefore actually nearly as small as they appear in the sky!). The Newtonian system intrudes upon God's space, with each and every star having its own solar system of planets.

51. Johannes Hevelius (1611–1687)

Sextant from *Firmamentum Sobiescianum*, Gdansk, 1687–1690
Engraving 12¼″ x 14¾″

52. Johannes Hevelius (1611–1687)

Virgo From *Firmamentum Sobiescianum*, Gdansk, 1687–1690
Engraving 12¼″ x 14¾″

53. Johannes Hevelius (1611–1687)

Aquila and Antinous from *Firmamentum Sobiescianum*, Gdansk, 1687–1690
Engraving 12⅜″ x 14⅞″

Hevelius, a wealthy and highly talented Danzig brewer, devoted his energies to building a major private observatory and to the pursuit of wide-ranging astronomical observations. He presented his findings to the world in a series of elaborate volumes for which he personally drew most of the illustrations. His last major project was this important star atlas, *Firmamentum Sobiescianum*, with constellation figures that he may have engraved himself. Most of the fifty-six copper engraved plates depict a single constellation or, occasionally, two or three smaller constellations. Like Bayer, he adjusted the scale of each chart to show the constellation to full advantage. But unlike Bayer, he viewed the sky as seen from the outside of a celestial globe. This rather old-fashioned orientation (used, for example, by Dürer) was rarely employed after Hevelius's effort.

Among the new constellations invented by Hevelius was Sextans Uraniae, modeled after one of his own splendidly decorated instruments. Its brightest star is of the faint fourth magnitude, so it is no wonder that the ancients had not made out any configuration in the area. Nevertheless, it provided Hevelius with the opportunity to immortalize one of his naked-eye measuring devices, which he had brought to an ultimate perfection of design and use.

Hevelius's *Virgo*, with her long, flowing tresses, her sheaf of grain, and her pointing hand, brings a new level of realism and artistry to this zodiacal depiction. Her carefully bordered gown and the finely detailed feathers of her wings add to the richness of the figure.

Hevelius's engraving of *Aquila and Antinous* dramatically accentuates a diagonal plan, common in High Baroque painting, yet not usually found in celestial cartography. Aquila is focused intently on the direction of the arrow to be shot by the handsome young Antinous. For context, Hevelius shows adjacent constellations in the background, with Sagitta plain and simple above the wonderfully shaded feathers of Aquila. Bevis's version of Sagitta (CAT. NO. 8) offers a clear example of how constellations viewed from Earth (within the celestial sphere) appear opposite to those viewed from Heaven. In the lower left of this image appears the constellation named in honor of the King of Poland, John III Sobiesci, to whom Hevelius's atlas is also dedicated. It was identified by the seven naked-eye stars on the shield and cross in 1684, three years prior to the printing of this atlas.

54. Johann Baptist Homann (1644–1724)

Schematismus Geographiae Mathematicae, Nuremberg, 1753
Engraving 18⅝″ x 21⅞″

Founder of a leading cartographic publishing house in Nuremberg, Johann Baptist Homann's name is often associated with the astronomical works of Doppelmayer (CAT. NOS. 42–44). In this engraving, Homann shows a rich collection of visualization aids for astronomical and geographical systems. The central location is given to Copernicus, with the solar system drawn to scale; below it is the visually pleasing system of Tycho represented in virtually all compendia of cosmology. The Compass Rose for terrestrial winds, and the Armillary Sphere depicting coordinates in the lower right, can be compared to Zatta's renditions in Venice twenty-six years later (CAT. NO. 81), and Lapie's versions (in CAT. NO. 59), published in France eighty-five years later. Homann offers simple, functional devices for teaching spatial relationships between Earth and sky.

55. Alexis Hubert Jaillot (seventeenth century)

Four Systems of Cosmology, Paris, c. 1690
Engraving 5¾″ x 8¼″

Four Systems of Cosmology shows a traditional depiction of the planetary systems of Ptolemy (geocentric), as well as Brahe (merged geo-heliocentric), and Descartes (Copernican). A final panel attempts to show the larger cosmological context of the Sun and fixed stars, with a curve of symmetry of the heavens. The clouds embellishing each panel appear to be purely decorative, similar to the style used by his contemporary in Paris, Nicolas de Fer (see CAT. NOS. 38–39). They do not suggest Descartes's whirlpool theory of planetary motion, as occurs in Buy de Mornas (CAT. NO. 17).

56. Alexander Jamieson (nineteenth century) with Neele & Son, engravers

Lyra, Hercules, Corona Borealis from *A Celestial Atlas*, published by G. & W. B. Whittakers, T. Cadell & N. Hailes, London, 1822
Engraving 8⅜″ x 10½″

57. Alexander Jamieson (nineteenth century) with Neele & Son, engravers

Ophiuchus, Serpens from *A Celestial Atlas*, published by G. & W. B. Whittakers, T. Cadell & N. Hailes, London, 1822
Engraving 8⅜″ x 10½″

Jamieson's 1822 publication of *A Celestial Atlas* offered an English version, in reduced size, of the great constellation figures of Johann Elert Bode (CAT. NOS. 12–14). The somewhat senior citizen versions of *Hercules* and *Ophiuchus* (holding *Serpens*) should be contrasted with the robust classical male figures in Coronelli's great globe gores (CAT. NOS. 32 & 33). The view is geocentric in Jamieson and external of the celestial sphere in Coronelli.

58. T. Jeffreys (eighteenth century)

The Geography of the Great Solar Eclipse of July 14, 1748, published by E. Cave, London, 1748
Engraving 12⅛″ x 17⅜″

To inform its citizens worldwide, the British Parliament published this chart in early 1748 showing how the solar eclipse of July 14th of that year would appear in various cities. The path of totality (labeled "Race of the Center") ran from Newfoundland to Celon. In the British Isles, maximum obscuration of the Sun by the Moon occurred near 10 A.M. (as shown for Dublin, Edinburg and London), a very civilized time for viewing. Note that this was a rare "annular eclipse" meaning that the Moon did not hide the full Sun at totality, but allowed a rim of light around the disk (as seen in Newcastle). Partial eclipses then progressed across Europe, the Middle East, and then through south India. In the American Colonies, represented by Quebec in Canada and Boston in New England, the eclipse was in progress at sunrise, and thus that particular summer day in North America got off to a late start.

The map upon which the eclipse path is shown reflects the latest results of geographical surveys at high latitudes, showing new coastlines discovered by G. Smith to its viewers, probably for the first time.

59. E. Lapie (eighteenth century) with Pierre-Alexandre Tardieu (1756–1844), engraver

Systèmes Planétaires from *Atlas Universel*, Paris, 1838
Engraving 20⅛″ x 26″

Systèmes Planétaires by the French geographer E. Lapie shows the planetary systems devised by Ptolemy, Copernicus, and Tycho Brahe, along with a plan depicting the full Universe, as it was known in the early nineteenth century, complete with the minor planets or asteroids. In the earlier systems the stars were thought to have been confined within a finite and fixed sphere, an idea outdated by Lapie's time. Other celestial phenomena, namely lunar and solar eclipses, the phases of the Moon, and an explanation of the seasons, are also noted. The armillary sphere in the center of the engraving represents the great circles of the sky, including the horizon, meridian, equator, tropics, polar circles, and ecliptic hoop. The circles are divided into degrees for angular measurement. Two cross sections of the sphere flank it on either side.

This engraving represents a very matter-of-fact approach to the study of astronomy. No figures or ornament decorate this diagram, showing how the restraint of the Enlightenment affected such representations.

60. Allain Manesson Mallet (1630–1706)

Eclipse of the Moon from *Description de l'Univers*, Paris, 1683
Engraving 5¾″ x 4″

61. Allain Manesson Mallet (1630–1706)

Phases of the Moon from *Description de l'Univers*, Paris, 1683
Engraving 5¾″ x 4″

62. Allain Manesson Mallet (1630–1706)

Ursa Major, Ursa Minor & Cetus from *Description de l'Univers*, Paris, 1683
Engraving 5¾″ x 4″

63. Allain Manesson Mallet (1630–1706)

Armillary Sphere from *Description de l'Univers*, Paris, 1683
Engraving 5⅞″ x 4¼″

64. Allain Manesson Mallet (1630–1706)
Types of Eclipses from *Description de l'Univers*, Paris, 1683
Engraving 5⅞″ x 4⅛″

65. Allain Manesson Mallet (1630–1706)
Copernican System from *Description de l'Univers*, Paris, 1683
Engraving 5⅞″ x 4⅛″

66. Allain Manesson Mallet (1630–1706)
Pole Arctique from *La Voye Laictée*, Paris, 1684
Engraving 6″ x 4¼″

67. Allain Manesson Mallet (1630–1706)
Pole Antarctique from *La Voye Laictée*, Paris, 1684
Engraving 6″ x 4¼″

68. Allain Manesson Mallet (1630–1706)
Members of the Solar System: Sun, Moon, Mercury, Venus, Mars, Jupiter and Saturn from *Description de l'Univers*, seventeenth century (German edition)
Seven Engravings 6⅛″ x 4¼″ each

Allain Manesson Mallet was a seventeenth-century French military engineer and later a teacher of mathematics to the pages of Louis XIV. His *Description de l'Univers* contains a curious mixture of information, including star maps, maps of the ancient and modern world, and a synopsis of the customs, religion, and government of each nation. Mallet himself drew most of the plans that were engraved for this work.

Several of the engravings from his book are included here. *Eclipse of the Moon* and *Phases of the Moon* (CAT. NOS. 60 & 61) have relegated their astronomical information to decorative curtains suspended in the sky. Beneath these ornate curtains charming scenes of harbor towns take on as much importance as the information conveyed above. The artist was clearly concerned with entertaining his readers while teaching them something about astronomy. Mallet's engravings contrast with those of Lapie and de Fer, whose plans are concerned almost exclusively with scientific information.

In *Ursa Major, Ursa Minor & Cetus*, well-fed animals depict northern sky constellations, while an instructive description of three-dimensional celestial coordinates appears in the *Armillary Sphere*. Aristotle's proof of the sphericity of the heavenly bodies is demonstrated in *Types of Eclipses*, in which a flat rectangular object is shown to cast a type of shadow never observed during eclipses.

The heliocentric theory is portrayed elegantly in *Copernican System*, and the constellations near the northern and southern celestial poles are depicted in *Pole Arctique* and *Pole Antarctique*.

Prior to Galileo's use of a telescope to capture the two-dimensional appearances of celestial objects, only the Sun and Moon had visual characteristics capable of being seen by the naked eye. Positional information dominated astronomical debate prior to 1610, and the previous eight images of Mallet all dealt with various aspects of organization and motion of "point sources" of reflected sunlight (i.e., the planets). Now, in *Members of the Solar System* from a German edition of Mallet's *Description de l'Univers*, the family of Sun, Moon, Mercury, Venus, Mars, Jupiter and Saturn all become real worlds. As is Mallet's style, the celestial bodies occupy a minor area in each image, with idyllic countrysides worthy of Turner for the Sun and Venus. Mercury is shown in crescent phase above a biblical landscape. The details for the Moon make it a physically meaningful place. The rough surface features of Mars come from the sketches of Huygens, the famous Dutch astronomer, perhaps accounting for the windmills below. Jupiter has its bands and the Medicean stars (i.e., its Galilean moons Io, Europa, Ganymede and Callisto). Saturn is prominent in the sky with its glorious rings. Taken together, they offer the reader a view of one's place in the physical Universe, from Earth to planets to stars. With Newtonian physics bursting upon the scene, "physical astronomy" would soon take hold to explore the natural causes of such diverse worlds.

69. Charles Theodore Middleton (eighteenth century)
Terrestrial Globe and Celestial Globe from *Middleton's Complete System of Geography*, London, 1750
Engraving 7½″ x 11¾″

Little is known about Charles Theodore Middleton, even though his *Complete System of Geography* was published in many editions throughout the eighteenth century. Matching geographical and celestial globes were prized possessions of kings, wealthy merchants and learned societies. While decorative and instructive to view, practicing astronomers relied more on flat maps where coordinate systems could be used readily for detailed positioning.

70. Julius Schiller (d. 1627)
Mary Magdalene from *Coelum Stellatum Christianum*, Augsburg, 1627
Engraving 12⅛″ x 16⅛″

71. Julius Schiller (d. 1627)
St. Helen from *Coelum Stellatum Christianum*, Augsburg, 1627
Engraving 12⅛″ x 16⅛″

72. Julius Schiller (d. 1627)
St. Joseph from *Coelum Stellatum Christianum*, Augsburg, 1627
Engraving 12⅛″ x 16⅛″

73. Julius Schiller (d. 1627)
Transitus Israel from *Coelum Stellatum Christianum*, Augsburg, 1627
Engraving 12 ⅛″ x 16 ⅛″

In the year of his death, Julius Schiller's masterwork, *Coelum Stellatum Christianum*, was published. A Bavarian lawyer like his friend and colleague Johann Bayer, Schiller maintained a lifelong interest in astronomy. *Coelum Stellatum Christianum* presented the most recent stellar information in a completely different context: newly conceived constellations drawn from Christian iconography rather than classical mythology. For the signs of the zodiac, Schiller substituted the twelve Apostles; for the mythological creatures of the northern hemisphere, Christian saints and symbols; and for the denizens of the southern hemisphere, heroes of the Old Testament.

The practice of matching classical personages with their scriptural counterparts had a long tradition in the Church, going back to Augustine. Many centuries earlier the Venerable Bede had made such a suggestion with regard to constellation figures. More recently, in Schiller's own lifetime, Wilhelm Schickard made a fledgling attempt to Christianize the heavens by introducing scriptural quotations into star charts. Schiller, a devout Roman Catholic, took the final step, resulting in a work of incredible labor.

The finished atlas contained up-to-date star placements and over forty new constellations. Redefining the constellations was an especially complicated task. In addition to drawing the figural boundaries to coincide (more or less) with the contours of the traditional constellations, Schiller also attempted to make a meaningful connection between the classical original and the religious replacement, in the Augustinian tradition. The results can be seen in four matching pairs chosen from Bayer's contemporary atlas (CAT. NOS. 2–5) and from Schiller's. Thus, naughty Cassiopeia becomes the repentant sinner, Mary Magdalene. St. Joseph, protector of the Virgin Mary, replaces Orion as the protector of the heavenly places. *Eridanus* is interpreted as *Transitus Israel*, the escape of the Israelites from the Red Sea. The transformation of *Cygnus* into *St. Helen* (who searched for the remnants of the true cross) appears to have been motivated by visual considerations primarily: the cross that Helen holds in her hands replicates the T-shape of the swan in flight.

Coelum Stellatum Christianum was a collaborative effort. Schiller had the help of Bayer and Schickard for astronomical purposes, and probably the counsel of the learned clergy with respect to theological content. Two Augsburg artist-engravers, Johann Mathias Kager (1575–1634) and Lucas Kilian (1579–1637), were involved in redrawing the constellations and engraving the plates.

Astronomers failed to adopt Schiller's scheme despite the extent of his labors, the accuracy of the star placements, and the imaginative care lavished on constellation designs. Today, we can appreciate Schiller's atlas as a testimony to his personal piety and to the spirituality of his age.

74. Unknown engraver, after John Flamsteed (1646–1719)

Andromède, Persée, le Triangle Paris (?), c. 1830
Engraving 7⅝″ x 9⅝″

The constellations of Andromeda, Perseus, and the triangle are the focal point of this engraving after Flamsteed. Perseus holds a sword and the decapitated head of the Gorgon Medusa, whose visage according to myth was so horrifying that it turned all who looked upon it to stone. In the course of his travels with the Gorgon head, Perseus flew over Aethiopia, where he saw the maiden Andromeda chained to a rock as a sacrifice for a sea monster. Andromeda had been placed there as punishment for her mother's boast of beauty superior to the Nereids. Perseus killed the sea monster, rescued Andromeda from her chains, and later married her.

Le Triangle has been given several identifications since ancient times. Among them are the Greek letter delta (Δ), the delta of the Nile, the three parts of Earth (Europe, Asia, and Africa), and the Trinity.

This nineteenth-century representation of Flamsteed's observations copies the Fortin edition of 1776 in size and figure design (CAT. NOS. 47–49). The background of Andromède, Persée, le Triangle is left blank in order to make the blue-tinted path of the Milky Way discernible. Missing are gilded stars and highlights featured in the earlier French edition. Yet, the quality of the engraving is superior, with Andromeda, the archetypal damsel in distress, still chained and bare breasted, resigned to being chased by Cetus, the sea monster.

75. Unknown engraver (eighteenth century)

The Path of the Comet of 1759 published in the *Gentleman's Magazine*, London, 1759
Engraving 5″ x 7⅛″

The *Gentleman's Magazine* offered its readers a mid-eighteenth-century version of public outreach in science. Edmund Halley (1656–1742), who observed a comet in 1682 and associated it with ones seen in 1531 and 1607, used Newton's gravitational theory to predict its return in an approximate seventy-six-year orbit. Seventeen years after his death, the comet that now bears his name appeared essentially on schedule, an emphatic confirmation of Newtonian celestial mechanics. Less dramatic is the simple journalistic cartoon used to aid Enlightenment magazine readers interested in locating the great comet.

76. Unknown engraver-cartographer (nineteenth century)

The Earth, Nebulae and Comets published by Blackie & Son London, c. 1880
Lithograph 9¾″ x 12¾″

The Blackie & Son print of 1880 is a lithograph (a print made from a plane surface on which the image to be printed is ink-receptive and the blank area ink-repellent). This method offers a visual texture that is particularly appropriate for the material depicted in the image. The subject matter offers a strange collection of items, a set that would not occur on the same page in a contemporary book on descriptive astronomy. The core of the image represents a traditional view of the Earth's motion about the Sun, not unlike Doppelmayer's *Phaenomena* (CAT. NO. 43). It is most surprising to see this linked with *Nebulae and Comets*, two of the most unlike astronomical objects known. Comets have always been a source of fascination to the public, and occasionally to astronomers as well. In the late nineteenth century, however, the great topic of debate concerned the distribution and nature of nebulae. Large clouds of interstellar gas, remnants of stellar explosions, clusters of stars and galaxies separate from our own were all called nebulae. Decades had to pass before the nebulous and cloud-like features associated with stars in our galaxy were distinguished scientifically from the external nebulae that ultimately proved to be the galaxies of an expanding Universe. Blackie & Son have used the realm of the nebulae in much the same way that earlier printers employed cherubs, banners, instruments, and scholars to fill the sides and corners around the central astronomical image. These fascinating objects are numbered and described, starting at the left (center) with the Crab Nebula, and proceeding clockwise to several of the great nebulae described by the well-known nineteenth century astronomer, Lord Rosse (1800–1867). In the lower left corner are several characterizations of Halley's comet as seen in 1835–36; in the lower right are comets observed during the early nineteenth century, prior to Halley's appearance.

The visual link that binds these unrelated objects is the blackness of space, whether simply the dark side of the Earth or the vast emptiness between spiral galaxies. Lithography of the late nineteenth century was well suited to express this condition.

77. Unknown artist (eighteenth century)

Constellations of the Zodiac from an unidentified Islamic book, Istanbul (?), 1776
Painting 10¾″ x 14⅝″

78. Unknown artist (eighteenth century)

Planispheres from an unidentified Islamic book, Istanbul (?), 1776
Painting 10⅜″ x 13⅞″

These two images are unique in the collection as the only representations of celestial phenomena from an Islamic source, and the only two items that are paintings, rather than prints. Each image is signed and dated by a different artist, with the same 1154 date in the Islamic calendar. To convert this date to the Western calendar system, the epoch year of the Islamic calendar system must be added, when Mohamed made his historic trip from Medina to Mecca in 622 A.D. Thus the paintings were created in 1776. The Ottoman Empire ruled Istanbul and beyond, and books were still mandated to be hand-transcribed with paintings for illustrations, apparently in an attempt by the Sultans to control the flow of information that printing presses might otherwise make accessible to the public.

In *Constellations of the Zodiac*, the classic twelve divisions of the Sun's path through the background of stars are shown, including the image of a human in the case of Aquarius (holding a water pitcher). However, Virgo (next to Libra) and Gemini are replaced by symbols. All labeling is in Arabic, both old and new styles. The terrestrial globe in the center is reasonably accurate for the period, but more representative of terrestrial geography than as an aid to navigation. Note the exaggerated number and sizes of islands in the Caribbean. The overall style is similar to the great zodiac plate of Cellarius (CAT. NO. 23). The two seals left and right are later additions since they refer to the Turkish Republic (started in 1924). Decorative embellishment representations of Earth, Air, Fire, and Water adorn each corner.

In *Planispheres*, the constellations of both hemispheres are shown with animals and human figures. Islamic practice traditionally avoided the representation of the human form, therefore the inclusion of these human figures with eastern faces shown both clothed and naked is most unusual. The painted figures are simple and lack the animation that can be found in the planisphere of Afferden (CAT. NO. 1) or de Fer (CAT. NO. 39). The difference between a painting and a printed rendition is clearly evident by the inclusion of only a simple ecliptic coordinate system (as also used by de Fer). The styles are as basic as shown in Walker (CAT. NO. 79) where minimal line drawings do not portray depth or three-dimensionality in the figures. Virgo is a placid woman with arms outstretched, while Andromeda appears male (Christ-like?) and Cassiopeia as an androgynous youth. The twins of Gemini are male and female, frontally naked, perhaps Adam and Eve before The Fall, as in paintings by Dürer and others. One is tempted to guess that Jesuit missionaries effected a revision of the traditional personages assigned to the heavens in order to accommodate both the Church and Islam, "adjusting" those that involved women of ill-repute, or young boys among the stars.

79. J. & C. Walker, drawn by W. Newton; constellations by W. Clarke (nineteenth century)

The Constellations: No. 3, Equinoctial and *The Constellations: No. 4, Solstitial* published by Baldwin & Cradock under the Superintendence of the Society for the Diffusion of Useful Knowledge, London, 1830
Engravings 12⅛″ x 12⅛″ each

These engravings of the constellations visible at the equinox and the solstice by the English engravers J. & C. Walker indicate the magnitudes of the stars with their classical and Islamic names. The figures of the constellations are classical in appearance and are painted in pale pastel colors. No attempt has been made, however, to indicate depth and three-dimensionality in the bodies, but rather the colors are used simply to differentiate objects and figures. This work is a very functional and practical guide to the skies. Such simplicity was a typical feature of the early nineteenth century. In this regard, the Walker engravings can be contrasted to the eighteenth-century English works by Hevelius and Bevis, where greater attention was paid to the modeling and coloring of the figures.

80. Antonio Zatta (eighteenth century)

Planisferio Celeste Settentrionale Tagliato Sull' Equatore from *Atlante Novissimo*, Venice, 1779
Engraving 12⅝″ x 16⅜″

81. Antonio Zatta (eighteenth century)

Tavola Sferica from *Atlante Novissimo*, Venice, 1779
Engraving 12⅝″ x 16⅜″

82. Antonio Zatta (eighteenth century)

Posizione Diversa Degli Abitanti Della Terra, Venice, 1779
Engraving 12¾″ x 16¼″

Some celestial maps were intended not so much for astronomers as for inclusion in terrestrial atlases. *Planisferio Celeste*, a star map of the northern sky, follows the format, style and content of constellation maps from Hevelius (CAT. NOS. 51–53). In each corner is one of the great observatories of northern Italian universities (Padua, Pisa, Milan and Bologna).

Tavola Sferica, an austere instructive description of celestial coordinate systems and terrestrial wind directions, represents the linkage between Earth and sky included in many atlases of geographical maps. Zatta was a prolific producer of atlases and maps for the maritime republic of Venice. His work formed the basis of more decorative versions of the same information, as shown in a G.M. Cassini engraving (CAT. NO. 22) published in Rome a decade later.

In *Posizione Diversa Degli Abitanti Della Terra*, a multipurpose chart, Zatta informs the viewer of an amazing variety of positional details relevant to geophysics and astronomy. The inhabitants of Earth are shown in various latitude zones, ordered by gravity along the limb (edge), and by civilization along latitude lines, with women and men in contemporary dress. Below is a demonstration of the bending of starlight (refraction) caused by the Earth's atmosphere. Filling the gaps top-bottom, left right, are images of the Sun and planets, each with a specific diameter (in kilometers) shown to make it clear that the engravings are not to scale. The moons of Jupiter and Saturn are shown displaced radially from each planet, in contrast to earlier depictions that tended to show them all at the same distance from their parent body (as in Cellarius, CAT. NO. 23, a century earlier).

Boston University
President *ad interim*: Aram V. Chobanian, M.D.
Provost *ad interim*: David K. Campbell

College of Arts and Sciences
Dean: Jeffrey Henderson
Chair, Art History Department: Kim Sichel

College of Fine Arts
Dean *ad interim*: Walt C. Meissner
Director, School of Visual Arts: Judith L. Simpson

Boston University Art Gallery
855 Commonwealth Avenue
Boston, Massachusetts 02215
617/353-3329
www.bu.edu/ART

Director & Curator: Stacey McCarroll
Assistant Director: Rebekah Pierson
Security: Evelyn Cohen
Gallery Assistants: Ross Barrett, Joshua Buckno, Sarah Dillon, Maree Emberton, Joshua Goode, Stephanie Inagaki, Meghan Lantzy, Rebecca Love, Carrie Moore, Dushko Petrovich, Natania Remba, and Gabriel Sosa.

Design and printing by The Stinehour Press
Lunenburg, Vermont

Bound by Acme Bookbinding, Charlestown, Massachusetts